WORLD MODELING: A DIALOGUE

NORTH-HOLLAND/TIMS STUDIES IN THE MANAGEMENT SCIENCES

Editor in Chief

MARTIN K. STARR

Volume 2

NORTH-HOLLAND PUBLISHING COMPANY – AMSTERDAM • OXFORD
AMERICAN ELSEVIER PUBLISHING COMPANY, INC. – NEW YORK

WORLD MODELING: A DIALOGUE

Editors

C. WEST CHURCHMAN

and

RICHARD O. MASON

1976

NORTH-HOLLAND PUBLISHING COMPANY – AMSTERDAM • OXFORD
AMERICAN ELSEVIER PUBLISHING COMPANY, INC. – NEW YORK

This North-Holland/Tims series is a continuation of The Professional Series in The Management Sciences, *edited by Martin K. Starr.*

North-Holland ISBN for the series: 0 7204 2450 X
North-Holland ISBN for this volume: 0 7204 0388 X
American Elsevier ISBN: 0 444 11029 1

Published by:

North-Holland Publishing Company – Amsterdam
North-Holland Publishing Company, Ltd. – Oxford

Distributors for the U.S.A. and Canada:

American Elsevier Publishing Company, Inc.
52 Vanderbilt Avenue
New York, N.Y. 10017

Library of Congress Cataloging in Publication Data

Main entry under title:

World modeling : a dialogue.

(North-Holland/TIMS Studies in management sciences ; v.2)
1. Economic history--1945 --Mathematical models--
Addresses,essays,lectures. 2. Social history
1945 --Mathematical models--Addresses,essays,lectures.
I. Churchman, Charles West,1913 II. Mason,Richard O.
HC59.W66 1976 330.9'04 75-44206
ISBN 0-444-11029-1 (American Elsevier)

Printed in the USA

CONTENTS

GROWTH SCIENCE – 1974
(COGITAMUS ERGO NIHIL)

i Society writes its own rules
but slowly
We rulers use our arrays
to fight ourselves
Paying the penalty for our persons
we drift into soggy snowbanks
Grayness surrounds our decisions
and we move in no sensible directions

ii Rats lice and mice
shit
We can all rest
assured that
One man's output is
another man's input
Equality is equivalent to evacuation
on a grand scale

iii Nature makes its own work
for all
We suffer little children
to process the results
Memories fuel the air creating
negative energy
Ferns pickaxed run to completion
losing faith

iv Squandered energy spawns showers
of prideful sparks
Making the city a home
for eyeless heads
Floralized spitoons/inverted cuspidors
pour spitchcock
On pygmalions fashioning themselves
into miniature gods

v

Crossing a painted desert
we are shadowed against purple cliffs
Painfully pushing the sand
we wander in wide circles
To avoid crossing
the lethal river
Ours if the foothill of knowledge
and the fight against time

vi

Serrated high rises are
ready to snare
A hawk already trapped
and seduced in his line of flight
Points of absolute light shadow
a stochastic dip in an ergodic sea
For organisms replenishing sunshine
with moonbeams

vii

We have a place to go and
we'd like to get there alive
Claiming a personal debt
we proudly prod the scraps
After hymns have been sung
to the evil of strange ways
What talisman stands against
inevitable decay

David B. HERTZ

NOTES ABOUT AUTHORS

Sumer C. Aggarwal ("Possible New Taxation for Environmental Improvement of the World") is Professor of Management Science, The Pennsylvania State University. He got his education in India, Russia, and USA. He has published two books and a large number of his papers have appeared in US and foreign journals. He is a member of TIMS and an approved Professional Engineer. He is consultant to a few organizations and frequently speaks in Executive Seminars and Management Programs. His current research interests are in the areas of Inventory-Distribution Systems, and Systems Analysis of Operating Systems.

Frederick Betz ("Structural Global Models") is Associate Professor of Statistics and Operations Research at the Wharton School, University of Pennsylvania. He received a B.S. from the University of Chicago and a Ph.D. in physics from the University of California, Berkeley. In the past, Professor Betz held academic posts at California State University, Hayward, and the State University of New York, Buffalo. His publications have appeared in several journals. His interest generally is in the management of social forms in which technology is developed and applied to problems of society. He writes in these two related areas: (1) science policy, administration of higher education, management of research and development, and (2) methodology of the social sciences and the construction and use of formal models in the social sciences. He is a member of AAAS and TIMS.

Edward S. Boylan ("On Some Policy Issues Raised by The Limits to Growth") is an Associate Professor of Mathematics at Rutgers University, Newark College of Arts and Sciences. He is also a consultant to the Hudson Institute, Croton-on-Hudson, New York, where he first became interested in policy issues in general and system dynamics in particular. He holds an A.B. from Columbia College and a Ph.D. in Mathematics from Princeton University. Previous papers have appeared in a number of journals including Econometrica and Management Science. He is a member of TIMS and the American Mathematical Society. Professor Boylan is currently on leave at Columbia University where he is a Visiting Scholar with the Department of Mathematical Statistics.

Kenneth R. Britting ("Backward Integration of System Dynamics Models – A Useful Validation Test?") is a Research Associate at the Alfred P. Sloan School of Management, Massachusetts Institute of Technology. He earned an Sc.D. from M.I.T. in control systems engineering. He is currently teaching and performing research in System Dynamics. Dr. Britting is the author of Inertial Navigation Systems Analysis, and Social System Modeling (in press). He is a member of Tau Beta Pi, Pi Tau Sigma, and Sigma Xi. His present address is M.I.T. E52-561, Cambridge, Mass. 02139, USA.

C. West Churchman ("The Niggling and the Grand: An Assessment of World Modeling") is now Professor of Business Administration and Research Philosopher at the Space Sciences Laboratory, University of California, Berkeley. He was editor-in-chief of Philosophy of Science from 1949 to 1959, and of Management Science from 1956 to 1961, president of The Institute of Management Sciences in 1962, and chairman of TIMS Council in 1963, as well as a member of the Council of

ORSA. He has served on the Public Health Service's National Advisory Allergy and Infectious Diseases Council, Bethesda, Maryland. For 1971–1972 he was appointed as Visiting Scholar to the Educational Testing Service, Princeton, New Jersey. Along with Ackoff and Arnoff, he published the first introductory text in operations research, Introduction to Operations Research. He wrote Theory of Experimental Inference, Methods of Inquiry, and Prediction and Optimal Decision, books that deal with the relationship between the philosophy of science and ethics. Challenge to Reason was published by McGraw-Hill in January, 1968, and was selected as one of the "Best Books in Management 1968" by the Academy of Management. The Systems Approach was also published in September, 1968 by Delacorte Press, New York, and received the McKinsey Book Award for one of the best books on management. His latest book is On the Design of Inquiring Systems, published by Basic Books in 1971.

Lieutenant Commander Rolf H. Clark ("Global Methodologies and Limited Analytic Techniques: A Proposed Merger to Attack Complex Problems") is attending the University of Massachusetts under sponsorship of the U.S. Navy's Doctoral Studies Program. His current research is in the field of managerial economics, and stresses defense manpower planning. Previous assignments include shipboard duty and management of a Naval research group engaged in military tactical analysis. He has lectured extensively on that subject and has received two Navy commendations for contributions in the field. He recently has published several articles relating to manpower planning and management science. He holds a B.S. from Yale, an M.S. in Operations Research from the Naval Postgraduate School, and expects to complete his current degree in 1975. He is a member of TIMS, ORSA, AIDS, and the AEA. Present residence is 114 Pleasant Street, Granby, Mass., USA.

José A. Costacurta de Azevedo ("Structural Global Models") is researcher at the Space Research Institute – INPE – in Brazil. He received a B.S. in Industrial Engineering from São Paulo Catholic University and M.S. in System Analysis from INPE's Division of Graduate Group. His M.S. dissertation was published in a book edited by INPE: System Engineering; Planning and Project Control. He is also co-author of some working papers published by INPE and National Research Council in Brazil. Presently he is working on his Ph.D. dissertation in the Social System Science Department at Wharton School (University of Pennsylvania). He is a member of SOBRAPO (Brazilian Society of Operation Research affiliated with IFORS).

Emilio Fontela ("Current Perceptions of the World Problematique") was Doctor in Economics (Geneva, 1962), visiting Professor at Case Institute of Technology (Cleveland, 1966), associate Professor of national accounting and econometric modelling at the University of Geneva (1970–1974), and is Professor at the University of Geneva since January 1975. He was manager of the Department of Applied Economics of Battelle-Geneva from 1966 to the end of 1973. While at Battelle he personally contributed to the development of large economic models (ACT, EXPLOR) and to the completion of many economic studies for private and public organisations throughout the world. Since January 1974, he has been consultant to Batelle-Geneva Management.

Author of numerous publications in the area of economic research, associate editor of the European Economic Review and of Futures, Professor E. Fontela is also vice-president of the European Association of Applied Economics (ASEPELT).

Jay W. Forrester ("Educational Implications of Responses to System Dynamic Models") is Germeshausen Professor, Massachusetts Institute of Technology, Cambridge, Massachusetts and is a member of the National Academy of Engineering, and a Fellow of the Institute of Electrical and Electronics Engineers, the Academy of Management, the American Academy of Arts and Sciences, and a Benjamin Franklin Fellow of the Royal Society of Arts (London). He is a director of the ALZA and Dynapol Corporations. Professor Forrester lived on a Nebraska cattle ranch until he studied at The University of Nebraska where he received the B.S. degree in Electrical Engineering in 1939. He received the S.M. degree from M.I.T. in 1945 and has been awarded honorary Doctorate degrees in engineering from The University of Nebraska, Newark College of Engineering, and the University of Notre Dame, and in science from Boston University and Union College. He received the Inventor of the Year Award from George Washington University (1968); the Valdemar Poulsen Gold Medal from the Danish Academy of Technical Sciences (1969); The Medal of Honor (1972) and the Systems, Man, and Cybernetics Society Award for Outstanding Accomplishment (1972), both from the Institute of Electrical and Electronics Engineers; the New England Award (1972) of the Engineering Societies of New England; and the Howard N. Potts Award (1974) from The Franklin Institute. He is author of Industrial Dynamics (M.I.T. Press, 1961), Urban Dynamics (M.I.T. Press, 1969), Principles of Systems (1968) and World Dynamics (1971; revised 1973) (both Wright-Allen Press, 238 Main Street, Cambridge, Mass.).

André Gabus ("Current Perceptions of the World Problematique") joined the Battelle staff in Geneva in 1964. Previously he worked as economic advisor to the general management of a large Swiss bank. He had practice in the distributive trades and food industry after he was graduated in economics from the University of Geneva in 1953. In charge of coordinating and developing socio-economic and econometric activities in Battelle-Geneva, André Gabus has worked closely with scientists engaged in research on such varied topics as multisectorial economic modeling, economic and social development, sociometrics, education, social indicators, simulation gaming, system analysis and finance. Much of his own effort in recent years has been in translating socio-economic research results into applications for long-range planning. His interest is concentrated on approaches making it possible to link techniques for systematic treating of human judgment to models with a view to bring simulation gaming closer to the reality of society. "Mutation" evaluation and goals identification feature his present research activities.

Olaf Helmer ("Interdisciplinary Modeling") received Ph.D. degrees in mathematics and logic from the Universities of Berlin and London respectively. He taught at the Universities of Chicago and Illinois, at New York City College, and at the New School for Social Research. In 1946 he joined Project Rand, which in 1948 became the Rand Corporation, of which he was a senior staff member of the Mathematics Department until 1968. During the academic year 1965–1966 he also held a half-time appointment as a Research Associate in the Institute of Government and Public Affairs at UCLA. He was one of the founders of the Institute for the Future, served as its President in 1970 and 1971, and later as its Director of Research. In 1973, he joined the faculty of the Graduate School of Business Administration, University of Southern California, as the Harold Quinton Professor of Futures Research.

In 1946, he participated in the design of the historically first systems analysis. He also was a co-designer of the Delphi and cross-impact methods. In recent years, his principal contributions have been in scientific methodology, in gaming and other simulation studies, and in long-range forecasting.

Among his publications are a book on Social Technology, which appeared in 1966 under the Basic Books imprint, and numerous articles, such as "On the Epistemology of the Inexact Sciences" (co-author: N. Rescher), Management Science 6 (1959); "Report on a Long-Range Forecasting Study" (co-author: T. Gordon), first published as a Rand Paper, P-2982 (1963), subsequently included as an appendix in Social Technology (see above) and also published in French, German, Italian, Japanese, Russian, and Swedish; "Planning Education for the Future" (co-authors: M. Adelson, M. Alkin, C. Carey), American Behavioral Scientist 10 (1967); "Analisi del Futuro, Il Metodo di Delfo", Rivista Italiana di Administrazione Industriale 9 (1967); "The Future of Science", Science Journal (October 1967); "Political Analysis of the Future", Institute for the

Future publication P-1 (1969); "On the Future of the Union", Institute for the Future report R-27 (1972); "Cross-Impact Gaming", Futures 4 (1972).

Erich Jantsch ("Modeling the Human World: Perspectives") is currently with the Center for Research in Management Science at the University of California, Berkeley, and a frequent visitor to various European universities. His books include Design for Evolution; Technological Planning and Social Futures and the widely translated Technological Forecasting in Perspective. His present work – focusses on nonequilibrium approaches to understanding the evolution of human systems. He is an Associate Editor of Management Science. Present address: 1962 University Avenue, Apt. 4, Berkeley, Calif., 94704, USA.

André Laurent ("Can We Model the Wild World or Should We First Tame It?") is an Associate Professor of Organizational Behavior at the European Institute of Business Administration (INSEAD) in Fontainebleau, France. He is a graduate of the Ecole de Psychologues Praticiens de Paris and holds a Licence in Sociology and a Doctorate in Psychology from the Sorbonne. Prior to joining INSEAD, Professor Laurent was Research Associate and Study Director at the Institute for Social Research, University of Michigan. He has published several articles on interviewing research.

Spyros Makridakis ("Can We Model the Wild World or Should We First Tame It?") is an Associate Professor at INSEAD (European Institute of Business Administration). He holds a diplome from the Graduate School of Industrial Studies in Greece and an M.B.A. and Ph.D. from N.Y.U. He has published widely with research interests in forecasting and general systems theory.

Richard O. Mason ("The Search for a World Model", and "A World Issue Debate: On Assumptions Underlying World Models") is Associate Professor in the Graduate School of Management and formerly Assistant Dean of the Professional Master's Program at the University of California, Los Angeles. He is currently on leave from U.C.L.A. and is Program Manager in the Division of Advanced Productivity Research and Technology, Research Applications Directorate at the National Science Foundation, Washington, D.C. He holds a Ph.D. in Business Administration from the University of California, Berkeley. He is a member of AAA, AAAS, ACM and TIMS.

Hasan Ozbekhan ("The Predicament of Mankind") is Professor of Social Systems Sciences at the Wharton School of the University of Pennsylvania. He was educated at the University of Paris and the London School of Economics. He is a member of the Editorial Boards of Futures (England) and of Technological Forecasting and Social Change. He also holds memberships in TIMS, ORSA, the North American Society for Corporate Planning and various other associations. Professor Ozbekhan is a consultant for planning to the French and Turkish governments, and has recently completed a large-scale research project on the International Functions of Paris, sponsored by the former. He is currently engaged in work in planning theory.

Michael H. Rothkopf ("Limits to Models") is currently a member of the scientific staff of the Xerox Palo Alto Research Center, Palo Alto, California 94304. He is the author of papers entitled "World Models Won't Work"; "An Economic Model of World Energy: 1900–2020"; and (with co-author Harry deVries) "Modeling Future Energy Supplies". In 1972 as head of the Planning Techniques and Economic Models Division of Shell International Petroleum Company, Ltd., he co-ordinated Shell's efforts to evaluate the Club of Rome Models. He has a B.A. degree in mathematics from Pomona College and an M.S. degree in Industrial Management and a Ph.D. degree in operations research from M.I.T. He has written papers on a variety of operations research topics including competitive bidding and scheduling theory. He is a member of the Operations Research Society of America and The Institute of Management Sciences.

Martin K. Starr ("The Logistics of Size") is Professor and Vice Dean at the Graduate School of Business at Columbia University. He received his B.S., Massachusetts Institute of Technology, 1948; M.S., Columbia University, 1951; Ph.D., Columbia University, 1953. Professor Starr is Editor-in-Chief of Management Science (Theory and Application sections) and President of The Institute of Management Sciences (September 1974–August 1975). His latest text co-authored with Irving Stein is The Practice of Management Science. He has written Systems Management of Operations, and Management: A Modern Approach. He has also written Product Design and Decision Theory, Production Management: Systems and Synthesis (2nd ed.) and with David W. Miller, Executive Decisions and Operations Research (2nd ed.), The Structure of Human Decisions, Inventory Control – Theory and Practice. Professor Starr has edited both Management of Production, and Executive Readings in Management Science. Professor Starr has published numerous papers in a variety of journals and has been chairman of many professional meetings for IFORS, TIMS, ORSA and the Society for General Systems Research. Professor Starr is a member of Beta Gamma Sigma, ORSA, TIMS, European Society for Opinion and Marketing Research.

Jan Tinbergen ("Development and Environmental Aims: An Intuitive View"), born 1903, took his Ph.D. at Leyden University in 1929. He was in charge of business cycle research until 1949 (1936–1938 at League of Nations) and then became director of the Central Planning Bureau of The Netherlands. Professor at Netherlands School of Economics, 1933–1973. Nobel memorial prize in economics, 1969.

Sir Geoffrey Vickers, VC ("Problems of Distribution") was born 1894 in England, graduated in the humanities at Oxford University, (1913–14 and 1919) where his studies were interrupted by four years service as an infantry officer in World War I. He then became a solicitor and practised as a corporation lawyer in London until 1940. During the next fifteen years, he was a public administrator, directing economic intelligence in World War II, followed by ten years as legal adviser to and member of the Board formed to organise the British Coal Industry. Since 1955 he has lectured and published many papers and six books on the application of systems theory to government and management and to contemporary problems. He has been a director of companies, has held many public and professional positions and is an honorary Fellow of the Royal College of Psychiatrists. His address is Little Mead, Manor Road, Goring, Reading, RG8 9ED, England.

Richard D. Wright ("Backwards Integration Tests of Dynamic Models") is Assistant Professor at the Department of Industrial and Systems Engineering, Georgia Tech. He received an S.B. and S.M. from M.I.T.'s Sloan School where he is pursuing a doctorate. Mr. Wright's current research interests are model validation and corporate simulation applications. He is a member of TIMS, AIDS, and Simulation Councils.

THE SEARCH FOR A WORLD MODEL

Richard O. MASON
Graduate School of Management,
University of California, Los Angeles

1. Medusa and Perseus

World issues have a Medusan quality about them. One recalls from the ancient myth that the Gorgon Medusa was a wretched woman, cursed with staring eyes, protruding fangs for teeth and writhing snakes for hair. Many of today's most pressing world problems seem much the same. Problems of the environment, growth, inflation and social structure are omnipresent, staring at us, challenging us to face them. But, they are every bit as entangled, ensnarled, complex and wicked as was Medusa. Today's world issues are difficult to formulate, dangerous to embrace, paralyzing to confront.

Aurelie Peccei, the Italian industrialist who founded the Club of Rome, refers to this situation as the *problematique.* This is his term for the cluster of intertwined problems which together produce "worldwide symptoms of a general but as yet little understood malaise." Our most crucial task today, he says, is to understand it.

Several years ago Peccei and his colleagues set out to do something about the problematique. With a group of concerned world leaders they formed the Club of Rome and in turn sought to build a large-scale computer model which would provide insight into some of the world's problems. M.I.T. Professor Jay Forrester was asked to apply his "Systems Dynamics" concepts to the problems of unchecked growth in the world. The effort culminated in a report authored by Dennis Meadows entitled *Limits to Growth.*

Limits to Growth represents a fascinating chapter in the evolution of management science. Few scientific reports since Darwin's *Origin of the Species* have stirred so much popular debate and certainly few, if any, management science efforts have ever created such widespread interest.

There are, of course, many who question the results. Some argue that it takes a Malthusian worldview as given and thereby underestimates the role that technological progress and man's ability to innovate might play in solving world problems. Others argue that the model contains such a restricted set of variables and that it is so highly aggregated that any conclusions must, of necessity, be grossly oversimplified at best, totally misleading at worst. Still other experts in the field have challenged the choice and definition of specific variables and relationships. For example, the model's treatment of energy, resources, technology, prices and economic control have all come in for criticism from those who believed that other formulations would have been more appropriate. In short, the work has been quite controversial.

But the important point is that for once people cared. Despite the intricacies of the model and its complicated feedback loop diagrams, both the nonscientific and the scientific community alike read, listened and thought. Whatever one thinks about the method or its conclusion, *Limits to Growth* has become a way of thinking about the world's problems. With this popularity a new and rather awesome responsibility comes to management scientists. The public has seldom paid so much attention to the products of our profession. More importantly, perhaps, world modeling efforts test the very limits and the foundations of management science.

A world model is the broadest possible form of modeling. All models are merely representations of the substantive areas being studied. In the modeling process the modeler is obligated to relate his model to the real and substantive area being modeled. Though it often goes unsaid, in large part, it is primarily the modeler's professional judgment which guarantees the accuracy and validity of the model. These kinds of judgments are debatable and always difficult to make. Developing a world model is the most difficult task a modeler can undertake.

The complications of creating a world model become apparent when we consider how relatively easy it is to model smaller systems. If, for example, the scientist wishes to model a firm's inventory he can visit the warehouse, talk to the people who are involved and, with some thought, he can decide what things to include in the model and what things to exclude. In other words he can "bound" the problem. Moreover, certain constraints and relationships will emerge as mandatory. These will be constraints, such as the capacity of the warehouse, which any reasonable observer would agree must be incorporated in the model.

Such is not the case with world models. A world model takes the entire world as its substantive domain and there are few if any mandatory constraints which the modeler must take into account. So the world modeler in a very real way has almost absolute freedom in the way he constructs his model. True, all of the world's citizens are bona fide observers of the substantive domain and therefore may serve as auditors of the model. But, each is just one observer among many and his perspective is limited. So, in the final analysis, the world modeler exercises the highest degree of judgment in the construction of his model. He thereby must accept the greatest amount of responsibility for the model and its uses. This is a moral obligation which should not be taken lightly.

The purpose of a world model is to provide policy makers with an organized representation of the world, a sort of a reflection of reality to aid their decision making. In the complex world of today policy makers need these images to guide their thinking. It is physically and psychologically impossible for them to confront the full complexity of the world without the aid of such imagery. So, despite their many shortcomings, world models are an absolute necessity for the world's policy makers.

One recalls that in the myth no one could not face Medusa directly without turning into stone, just as many policy makers today are incapable of facing the world's problems head-on. Perseus, of course, had a solution for this threat of paralysis in dealing with Medusa. He dared not face her directly for he, too, would turn into stone. Instead, he used a shiny shield, a gift from Athena, as a mirror. With the help of its -reflection he was able to track her and eventually to slay her. Modern day Perseuses also need shields if they are to attack the world's problems successfully. Models such as the *Limits to Growth* model are certainly good candidates. They cast interesting images of the world. But as we have seen world modeling is a very tricky business. There are many ways to go about it and there are many people whose professional judgment on modeling the world and its problems should be heard. In the succeeding essays we shall listen to a few of them and encounter the "shields" they offer as gifts to future slayers of world problems. But first some thoughts on choosing world models.

2. Choice of a shield

The notion that there is more than one way of modeling the world naturally raises the question of how policy makers should go about choosing from among the many possible models to guide them. There are several very general criteria they might employ. Briefly these are:

A model should organize existing data in such a way that it provides new insights and facilitates interpretation of the data.

A model should also serve the heuristic function of posing new questions and calling for new facts.

A model should permit its user to make important, verifiable predictions.

A model should provide the basis for collecting accurate measurements.

The *Limits to Growth* model stacks up pretty well by these general criteria; but, then so do other potential world models. These general criteria alone will most likely not suffice for policy makers to discriminate among competing world models.

Two other criteria, prominent in the history of science, suggest themselves. One is originality. How unique are the ideas emerging from the model? *Limits to Growth* fares reasonably well by this criterion. It is perhaps one of the most original and ambitious studies of its type ever attempted in the social sciences. (Although one ought to read Pavitt's provocative essay entitled "Malthus and other economists – some doomsdays revisited" [1] before arriving at his own conclusion.)

The other criterion is simplicity. The model should be a useful simplification of the substantive problem area. Here there are several notions of simplicity to apply. There is the economic notion of the efficiency of the model. Do the benefits derived from its use outweigh the costs of its construction and operation? Irrespective of questions of ultimate validity, it seems clear that the *Limits* model has generated enough debate and concern to justify its costs. We must credit it with being a worthy effort and not fault it here.

Simplicity has a different connotation, according to Copernicus. It involves minimizing the number of steps required to obtain a solution. The Systems Dynamics computer routines employed in the *Limits* study seem to satisfy this need. Again it survives the test.

Francis Bacon, on the other hand, rooted simplicity in one's everyday experiences. He advocated that a model must reflect the reality of the situation as the model users see it. It must be consistent with the world as they understand it. All world models, as we have discussed previously, suffer by this requirement. Although one might argue that the concepts and feedback loops used in *Limits* are not the most natural way of thinking about the problem, it is difficult to fault the *Limits* model on this score. It is, probably, as understandable in terms of everyday concepts and language as any. But the reader can make his own judgment on this score for everyone is an expert in terms of his own everyday experience.

There is one more approach to simplicity, however, a notion deriving from William of Occam, that we ought to take seriously. Occam's famous razor judges that model best which contains the least number of assumptions. Couple this requirement with the need for the assumptions to be realistic also, and a new line of questioning *Limits* is called for.

At least two kinds of assumptions underlie any model – formal and world view. The formal assumptions of the *Limits* model are revealed in the variables selected – resources, population, agriculture, capital, pollution, etc. –, their definitions, and the method of relating them via feedback relations. Formally the model's assumptions are both parsimonious and debatable. Many people in this volume and others have challenged the use of such a restricted set of assumptions. Others violently opposed the particular formulation employed.

It is when we come to the world-view assumptions, however, that the controversy heightens. The Malthusian and Ricardian overtones have been questioned by those who favor technological optimism. So was the emphasis on physical rather than social and political phenomena. These critiques and others speak to some of the historical values incorporated in the model. However, one might go further and also ask about the metaphysical values which underlie the model.

Limits assumes the world is a machine. Stripped to its essentials the world of *Limits* is one in which all things, events and processes are reducible to changes in the location of space-time elements. Once the machine is "started" it runs on by itself in an infinite series of cause and effect events just as though it were some sort of perpetual motion device. The theory of differential equations that underlies the Systems Dynamics method permits the machine to be a complicated one incorporating thresholds and feedbacks; but, it is a machine nevertheless. If important policy is to be based on this mechanistic model one ought to not only ask questions about the nature of the machine and suggest alternative machine designs but also ask whether the world is a machine at all. In choosing a shield we should speculate on other possibilities. Where are they to be found?

We have seen that ultimately there is no objective way to assess world models.

We must therefore rely on the simplicity of their basic assumptions and of their root metaphors if we are to choose from among them. These judgments of simplicity and relevancy are generally made in the context of opposition where the assumptions of one model can be compared with those of another. Our purpose in this volume is to provide sufficient variety and opposition of thought to sharpen our judgment on modeling world issues. Consequently, we have constructed a dialogue. We have asked various world modelers to describe the approach they would use in modeling the world. The result is a set of images and summaries of different perspectives which we hope will help policy makers to decide on the choice of a shield.

3. Images

The dialogue begins with Hasan Ozbekhan who describes the events leading up to the writing of the "prospectus" for the Club of Rome and presents his original concept of the "problematique". Ozbekhan argues that world problems have become so complex and intertwined that traditional reductionist and positivist methodologies are no longer appropriate. He calls for new, holistic world models and sketches out the course they might take. Fundamental to any world model is the worldview or value base that underlies it. He concludes that "ecological balance" should be the basic underlying value of any world model.

Forrester continues next by categorizing and then refuting the major criticisms of the Forrester–Meadows work. The practice of dynamic modeling, he argues, requires the modeler to focus on the dynamic properties of the system rather than its institutional character. The dynamic nature of the system should guide the modeler's choice of elements, relationships, levels of aggregation, time dependency and the basic data they imply.

Thus, he takes exception with the critics who begin with assumptions about the market structure and therefore shudder at the lack of a price mechanism in the *Limits to Growth* model. Likewise, he counters those who criticize the aggregation of scientific and technological innovation into a single variable. They are dynamically similar he maintains. The treatment of developed and underdeveloped countries, the role of time and delays or lags, and, indeed, the basic logico-hypothetical nature of the model are defended on the basis of dynamic theory. The argument reveals some of the key underlying assumptions in the dynamic systems world view and sets the stage for other points of view.

Both Ozbekhan and Forrester call for a critical review of the theory of modeling. Betz and Azevedo begin such a critique. Drawing on the philosophy of E.A. Singer they distinguish between cause-effect models and producer-product models. The latter include the willful actions of humans and their institutions as an integral part of the model. In arguing for more models using producer-product concepts Betz and Azevedo conclude that a network of regional models, aligned according to nation states, industrial organizations or other action taking institutional forms is preferable to a single global model since "there exists no global agency which has the ability to affect the whole world at once."

The producer-product model suggested by Betz and Azevedo raises the question of what assumptions a world model should include about the willfulness and planning ability of man. The next three authors deal with this problem directly.

For Vickers the power of man to affect the future, that is to engage in self-directed processes of change, is fundamental. Politics becomes his basic metaphor. For Vickers a world model should begin with an analysis of political structure. Consequently, he focuses on the nation states and assesses their degree of technological development and their degree of self-sufficiency. This review points to problems of political instability. He concludes that the cause of internal and international instability is the "failure to solve problems not of production but of distribution." Distribution problems will persist whether conservationist or expansionist policies are followed. Market structures alone are incapable of solving the problem. So, the mechanism by which the world's wealth is distributed becomes the starting point for Vickers' world model.

Politics is also the basis of Tinbergen's critique. This viewpoint leads him to make two important distinctions. With Vickers he believes that the tension between the developed and the underdeveloped and developing countries will increase and therefore must be accommodated in any world model. Secondly, he points out that there is a critical difference between a *forecast* (such as *Limits to Growth*) and a *plan*. Plans are man's efforts to control his destiny. Tinbergen then explores the implications of a plan (one that he thinks is likely to materialize) which is based on an average annual rate of a five percent increase in per capita income in the developing countries. He concludes that zero growth is possible but only with difficulty and only with active government participation in the development and execution of the plan.

Assumptions about motivation and behavioral modification are at the core of Aggariwal's proposal. He acknowledges the threat of a crisis in pollution and agriculture as predicted by *Limits to Growth*. But, with Tinbergen he believes that well conceived plans can mitigate against dire forecasts. This leads him to advocate a system of environmental and agriculture improvement taxes. He argues that a carefully designed tax policy will encourage people to behave in ways which are conducive to world survival. Thus, while Vickers and Tinbergen focus on nation states, Aggariwal starts with the individual and the enterprise in constructing his world model.

Contrary to the more social emphasis of Vickers, Tinbergen and Aggariwal, technological necessity and, indeed, technological optimism is Boylan's basic metaphor. The world is characterized as a race between man and technology. Boylan believes that *Limits to Growth* is pessimistic in that it fails to give adequate consideration to the technological possibilities of discovering new natural resources, creating new energy sources and extending the "green revolution". So his world model is rooted in technology.

With Helmer the focus changes. While the preceding authors' critiques concern the substantive assumptions which underlie world models, Helmer's critique is methodological in nature. He proposes five criteria by which world models might be judged. Econometric models, generally speaking, fail by all five. Systems Dynamics models, on the other hand, fail by only two. Systems Dynamics models do not fully satisfy the criteria because they are not probabilistic (which can be remedied) and they only deal with trends rather than events. In Helmer's world view uncontrolled (or at least unmodeled) exogenous events which mutually impact on one another are major determinants of the future world order. He advocates, therefore, a cross-impact methodology which accommodates both events and trends as the best method for modeling world issues.

Although their methodological approach is similar to the cross-impact analysis of Helmer, for Fontela and Gabus it is problems rather than events that form the basic unit for studying world issues. Their DEMATEL method involves constructing a matrix of problem interactions from the responses of informed observers of the world scene. The matrix can be analyzed to identify key problem sources (main dispatchers) and key problem destinations (main receivers or objectives). A graph of problem clusters is also generated.

The DEMATEL approach also raises another critical issue in world modeling. Namely, what is the most reliable source of information? *Limits to Growth* sought the most objective means of acquiring the data upon which the model is based. Helmer acknowledges the inexact nature of the science and explicitly seeks expert opinion by means of Delphi and other devices. The Fontela and Gabus model goes further. It is entirely based on perceptions. DEMATEL merely gives form to the perceptions. While Forrester and Meadows focus on the physical world, Fontela and Gabus focus on the psychological world. The result is a graph relating what Jung would have called the "collective unconscious". The graph depicts some 30 respondents' collective perception about important world problems.

Erich Jantsch makes this subjective/objective conflict even more explicit. He constructs a 3 X 3 taxonomy of world models based on two primary differentia: the type of reflection and the state of the subject/object relationship. *Limits to Growth* is essentially a mechanistic (statistically preprogrammed) model in his scheme because the modeler stands outside the system being modeled and reflects on its structure by means of perception. Jantsch's taxonomy suggests that we might also think in terms of an approach whereby the modeler assumes himself to be part of the system being modeled and employs a self-reflective type of consciousness. This he calls creativity.

If one accepts creativity as a goal in modeling how is it to be achieved? Mason's article offers one approach. The assumption is that creativity can be stimulated by well-structured debate. An earth resources debate based on an Hegelian dialectical model is presented. Its purpose is to expose underlying assumptions about our modes of conceiving world problems in order to encourage the creation of new world views.

The earth resource debate is inconclusive. It leaves the reader with the task of forming his or her own world view. Is man able to formulate these new world views?

Or, as Fauchaux, Laurent and Makridakis put it, "Can we model the wild world or should we first tame it?". Their answer to the question is that we must tame it first. This requires theories of political power, attention to values and an acknowledgement of constraints. These constraints are social and cultural as well as natural. Their exposition inevitably entails conflict in thought. The essential task of the world modelers in their view is to resolve these conflicts and to be explicit as to "what we *want* to become, what we *could* become and what we *should* become". A substantial challenge for world modelers!

Faucheux and his colleagues as have others criticize the Malthusian overtones inherent in the *Limits to Growth* model. Martin Starr takes his cue from Malthus but asks a different question. He is concerned with finding the optimal organization size to balance linear and geometric growth. Consequently, organizations are the underlying metaphor in Starr's world view. The delivery of the world's goods and services is mediated through various organizational forms. The organization's production efficiency, marketing costs and management costs all serve to determine its minimal cost output level and, hence, its optimal size S. Starr suggests that we could reduce total social costs and thereby attenuate the problems presented in *Limits to Growth* if all organizations operated at or near their respective S. Drawing on biological theory he develops an approach for discovering S. The theory may prove to be useful in designing and regulating organizations.

Two additional papers, one by Wright and another by Britting, explore the more technical nature of the "machine" which underlies the Systems Dynamics approach. If a cause and effect machine runs well in a forward direction one would expect that it would run equally well backward. However, attempts at backcasting or retrodiction using the *Limits to Growth* model have generated some perplexing results – world conditions prior to 1900 do not match historical data. This seems to be due to round-off error and to the method of numerical integration used.

Wright summarizes the source of these problems in the *Limits to Growth* context and points out that in retrodiction a stable negative feedback loop can be transformed into an explosive positive feedback loop. Yet, he observes, in real social systems positive feedback loops are seldom permitted to continue unabated. People change the nature of the machine when necessary.

Britting develops the conditions under which reliable retrodiction is possible. He concludes that "A properly formulated model is constructed to be 'valid' within a particular time span of interest". He leaves unanswered the question as to how one might determine an appropriate time span for a world model.

The point that people may change the functioning of the machine when necessary is also taken up by Clark. He argues that systems dynamics models such as those used in the *Limits to Growth* can be supplemented by decision theory models and linear programming models. His approach permits decision models to be embeded in world models.

In *Limits* to *Models* Michael Rothkopf presents some important considerations for model builders and users to take into account when thinking about world models. He urges us to recognize the limitations of world modeling efforts and suggests that potential modelers should consider the tradeoffs between allocating social resources to build models or employing them to attack the problem more directly in nonanalytical ways. In the process, of course, he challenges the world view of the modelers themselves.

If there is a lesson to be learned from this dialogue it is that world models bear the imprint of the world view of their designers. Can it be said of a world model, as it was of Homer's Illiad, that it has "a great master's genuine stamp, and that stamp is *the grand style*"? Churchman poses this, the ultimate question of assessment, in the closing chapter of the dialogue.

4. Quest for Athena

In ancient Greek mythology Athena was the goddess of wisdom. She gave her shield to Perseus so that he might challenge Medusa and, the myth tells us, he was successful.

Policy makers and action takers in the world today desperately need better information and better models for dealing with their problems. They need the wisdom of a modern-day Athena to provide them with effective and useful insights. The *Limits to Growth* study of Forrester and Meadows is a highly visible candidate for the role. Many are turning to it for guidance today. But, as we have seen, there are also other potential Athenas offering other shields to be used. In the papers which follow several thoughtful observers have briefly summarized their ideas so that we might benefit from them. The papers can be read as though they were a conversation about useful ways of thinking about and modeling world issues. If the dialogue is successful we will have launched the quest for a new Athena and aided in her discovery.

Reference

[1] H.S.D. Cole, Christopher Freeman, Marie Tahoda and K.L.R. Pavitt, Models of Doom (Universe Books, New York, 1973).

THE PREDICAMENT OF MANKIND

Hasan OZBEKHAN
University of Pennsylvania

1. Introductory comments ... (*ex post!*)

A few weeks ago West Churchman suggested that, if suitably shortened and edited, the original "prospectus" I had written for the Club of Rome might be used as a preface for this issue of *Management Science* which is concerned with world modeling. I was, of course, happy to say "Yes!".

Only after I started to work on the shortening and editing did I begin to realize how much time had passed since 1970 when the original document was first circulated to most of the members of the Club of Rome. I mean internal time, not simply four relatively short years.

I experienced difficulties in trying to adapt the text to the purpose at hand. Attempts at shortening it were constantly balked by the desire to introduce new clarifications and ideas that I have developed since. Attempts at editing soon changed the tone and meaning of what I had originally intended to say. And I was unhappy with all these distortions. I finally concluded that one way out of the dilemma might be to write these "Introductory comments" where I could hope to lay before the reader, not so much the circumstances that had led to the writing of the original text as some of the intentions which had governed my personal commitment to the project as a whole. This has permitted me to condense only that part of the original prospectus which I still believe to have some interest for present-day readers. Similarly the editing, quite minimal though it is, can now be justified by the necessity to maintain continuity and relevance in the shortened text. Thus, those parts of the article which follow this introduction represent – hopefully without major distortions – what I had originally written.

The initial version of the prospectus of the Club of Rome, called "The predicament of mankind – Quest for structured responses to growing worldwide complexities and uncertainties", was divided into three main sections. The first, which described in brief the "predicament of mankind" was cast in the form of a Work Statement and Proposal. The second outlined the Conceptual Frame of the approach to the project that had been described; the third, a cluster of Annexes, further elaborated the aims and constitution of the Club of Rome, detailing ideas such as the "World Forum" and describing the composition of the "Work Group". Only the first part of this paper is presented here since the interactive "problems" which were discussed in it are still with us.

The entire document, repeatedly revised, altered, and fine-tuned in collaboration with my colleagues on the Club of Rome's Executive Committee, had been produced for several reasons. We felt that the group that had called itself the Club of Rome needed a statement which clarified what it was about. We recognized the need for a logically consistent statement that would express the purpose and shared values that had brought us together in the face of worldwide dissonances whose accelerating rate of occurrence and massive dimensions had alarmed us. Lastly, we came increasingly to realize that the nature and complexity of the events we had undertaken to look at demanded something more than the mobilization of random goodwill. They demanded a quality of commitment that – beyond the individual dedication which existed in abundant quantities – was both systematic and sustained.

I then believed as I continue to believe now, that the main thrust of our effort had to be methodologically radical – creative in approach, inventive in design. I believed, and continue to believe, that the endeavor had to be long run, experimentally configured, open to and capable of ingesting paradoxes, pluralistic, evolutionary and – with luck – global in scope. I then felt and still feel, that such a seemingly ambitious view was in point of fact an imperative arising from the very nature of our situation; that we no longer had the freedom to choose the known, easier, more restricted, incremental approaches; that reductionist preferences still lurking in our minds or sensibilities would inevitably distort the meaning not only of what was surrounding us, but of our action, an action that was intended as the first – cautious and tentative, but nevertheless the first – step in a momentous confrontation. To put it another way, I then asserted and still assert, that if our chosen task was to understand a worldwide system of interactive, ill-structured "problems" and to figure out "structured responses," we had to impose upon ourselves the deep and rigorous discipline that a "scientific" approach signifies when that adjective is used in its highest and most precise meaning; namely, the attitude of being wholly respectful of the objects, facts, occurrences that are being inquired into, as well as of the purposes, values, aims and aspirations that motivate inquiry ... remembering that in the universe of events generated by human action, purposes, values, aims and aspirations are inextricable parts of objects, facts and occurrences and that such a conjunction is what we call "reality". And reality was the very subject of our proposed inquiry.

However, that reality (as I believe I have already indicated, by implication) appeared to contain such large-scale masses of events, such unexpected velocities of occurrence, such intricate patterns of causality, as to constitute a quite new, unexplored and unfathomed "situation". This perception allowed one to pre-suppose that neither current problem-solving techniques, nor the dicta of conventional wisdom which such techniques nurture, would be sufficient to deal with whatever it was that needed to be dealt with. The new situation quite explicitly required what T.S. Kuhn has called a "paradigm change;" that is a new world view sustained by a new *methodology*. I felt, therefore, that the effort contemplated by the

Club of Rome must first address itself, consciously and systematically, to the creation of such a methodology. It was this that I had in mind when I wrote: "*The governing statement concerning the project as a whole is that its aim is not research in the traditional sense, but 'invention'.* This should be understood to mean that what is expected from the effort are new insights and approaches rather than the further and deeper elaboration of already known facts. The latter will be used in their present state of elaboration as the substantive material upon which the work will bear – however, the expectation that animates the work itself concerns the *meaning* which all these facts, in their systemic nature and system-wide impacts, have for the future of mankind." (Emphasis in original text.)

Moreover, I saw this phase of the endeavor as a process that could be so designed that it would become something akin to what Russell Ackoff has since termed an "adaptive-learning system." For, if the universe of human events and human action is the subject, then the creation of a methodology to look into, or penetrate this subject is itself an act, a man-generated event – something which, by the very fact of being brought into existence, changes reality. It must, therefore, be iteratively re-invented, re-shaped or continually sharpened in order to remain viable.

Another aspect of this approach was that it demanded a critical review of orthodox methodology – a review undertaken to find out whether our intellectual instruments had become blunted with respect to the cutting edge that was now necessary; whether long usage had bent them out of shape; whether the simplifying algorithms they contained had limited their scope; whether, and in what ways, the new structure of reality might have rendered them either partially or wholly irrelevant.

I considered such an inquiry to be important because it is a precursor to the development of new "theories" – those rearrangements of values, outlooks and frameworks of concepts that imbue the data at hand with novel meaning and provide them with an organisation that explicates them more clearly than before. Such a new theoretic context would also legitimize the methodological innovations that might result from the proposed effort.

Many of the points I have touched upon in the preceding paragraphs have been further articulated in the works of Ackoff, Churchman, F.E. Emery, Trist and a few others including myself. It had then (in 1970) been my hope that the main thrust of the work to be carried out by the research group of the Club of Rome, of which group I was the Director, would, in its first phase, represent a massive effort in the area of methodology creation; that it would lead to new conceptions of large-scale modeling, perhaps of world modeling. That these expectations were not fulfilled and that the Club of Rome chose to explore other avenues of approach is a matter of record, or past history. What continues to be vitally important however is the predicament of mankind – the real deepening and darkening situation in which we all find ourselves.

2. A view of the world "problematique" . . . (*ex ante!*)

As in every epoch of its existence, mankind today finds itself in a particular "situation". And as always this situation is created and nurtured by those who live amid the myriad events that comprise it – events that are now in the process of tumultuous and ever accelerating change, events that now increasingly and even violently clash with one another. In some deep sense our situation compels us to animate and perpetuate it almost blindly, and thus to move toward a future whose shape or quality we do not comprehend, whose surprises we have not been able to fit into a coherent frame of ideas, whose complexities we are not in the least sure of being able to control.

There are, however, a few basic perceptions that possess both wide currency and increasing persuasiveness, by means of which people in many different walks of life have begun to apprehend the nature of this situation. Thanks to such perceptions we have come to recognize that the forces which hold us in their grip arise from the very source of our power and achievement – at least in those countries where the industrial mode of life has flourished – that extraordinary technological capital we have succeeded in accumulating and those all-pervasive analytic, positivistic methodologies which, by shaping our minds, have enabled us to do what we have done. More recently we have become aware that our achievement has, in some unforeseen (perhaps unforeseeable) manner, failed to satisfy those other requirements that would have permitted us to evolve in ways which, for want of a better word, we can call "balanced". It has failed to provide us with an ethos, a morality, ideals, institutions, a vision of man and of mankind and a politics which are in consonance with the way of life that has evolved as the expression of our success. Worse, it has failed to give us a global view from which we could begin to conceive the ethos, morality, ideals, institutions, and policies requisite to an interdependent world – this, despite the fact that the dynamics of our technologies and of our positivistic outlooks are global in their impacts, their consequences, their endless profusion and, more importantly, in the promises they proclaim and in the promises they imply.

This failure of vision is often regarded as having created a number of separate and discrete problems capable of being overcome by the kind of reductionist solutions our intellectual tradition can so readily generate. However, the experience of the past twenty or thirty years shows with remarkable clarity that the issues which confront us in the immediate present, as well as their undecipherable consequences over time, may not too easily yield to the methods we have employed with such success in the bending of nature to our will. Such apparent recalcitrance could be attributable to many things. It could be due to the magnification of the problems we must grapple with – that is, to the fact that almost all of them are worldwide in scope, whereas the sociopolitical arrangements we have created are ill-equipped for dealing with issues that fall outside their strictly bounded jurisdictions. It could be due to heightened yet often obscure interactivity among phenomena, whereas the efficiency of our manner of solving problems relies on the identification of clear and direct lines of causality. It may be due to rapid rates of change, especially in the technological sector, whereas our institutions, outlooks and minds are geared by long-time habit to beliefs in slow unfolding and permanence – beliefs which have sustained certain relatively stable concepts of polity, of social structure and of intellectual orderliness. While such discrepancies may in part account for our inability to cope with present issues, they also tend to exacerbate those issues. Few can question that the conjuncture of events – the situation – that surrounds us is to all evidence global, complex, dynamic, and dangerous.

Is is also new. Up to now our long commitment to stability and continuity has led us more or less successfully to steer our social evolution toward the known and to avoid that which, for being unknown, was also uncertain, and, therefore, frightening.

Today, confronted with dissonances that we can no longer ignore and fearsome uncertainties, our attitudes toward life and issues are tending to become rigidly polarized, and our actions hence hesitant to the point of paralysis. On the one hand we take refuge in the comforts of inertia, disguised as conventional social and economic remedies, which we believe will help us preserve what we have come to call and accept as "civilization". On the other hand we seek escape in iconoclastic or utopian futurisms whose feasibility, intellectual worth and operational efficacy we know to be questionable, but in whose visionary invocation of a wholly new human order we sometimes find solace as well as a fleeting sense of release.

These attitudes toward uncertainty are old. But, again, the uncertainty as we experience it today is new – both in its dimensions and in its extraordinarily complex dynamics and structure. Between the situation, whose general configuration most of us are to some degree aware of, and our manner of describing and assessing it there is a profound mismatch. And although many of us are aware of this mismatch, we yet go on describing the dissonant elements in our situation as singular "problems" and believing that such problems have correspondent "solutions". We do so not only out of fear and unfamiliarity but because it is in the nature of our languages, therefore of our manner of perceiving reality, to view difficulties as things that are clearly defined and discrete and to believe that they can be resolved if we manipulate them by means of some problem-solving technique. While certain problems (mostly in the fields of technology and engineering) can be dealt with in this way, it is quite apparent that such problems are no longer the most important ones.

The insufficiency of our notions of "problem" and "solution" becomes evident when we consider the truly critical issues that our situation proposes to both our intellect and to our conscience. Issues such as environmental deterioration, resource wastage, poverty, endemic ill health, hunger, urban blight, criminality, racial injustice, etc., clearly cannot be viewed as problems that exist in isolation – or, as problems capable of being solved *in their own terms.* Nor are they contained within specific or specifiable boundaries. Even the most cursory examination will reveal the more obvious (though not necessarily the most important) links between problems. Where endemic ill-health exists, neither poverty nor hunger can easily be divorced from it and vice versa. And the impacts of these conditions are not – cannot be – confined by regional or national frontiers. Environmental deterioration is as linked to resource wastage as to pollution, and the impacts of all three are global. Furthermore, if we try to solve any such problems exclusively in their own terms we quickly discover that what we take to be the solution of one category of problems may itself generate problems of another category (the reduction of death rates in developing areas and the resultant increase in poverty, urban blight, population growth, public unrest, etc., is a good example of this approach).

Another unfortunate consequence of our reliance on orthodox approaches is the misapplication of effort and energy. Many agronomists devote a great deal of ingenuity toward increasing the yield per acre of our crops without seeming to consider that the particular solution called "agriculture" may possibly no longer represent the single, feasible resolution of the problems clustered under rubrics such as "hunger" or "malnutrition" when the latter are considered in their worldwide dimensions. It seems reasonable, therefore, to postulate that the fragmentation of reality into closed and well-bounded problems creates a new problem whose solution is clearly beyond the scope of the techniques we customarily employ.

How, then, can we more realistically "see" reality? If we accept the interlinkage, interaction and dynamics of the problems that are proliferating in our situation it seems that we might approach a clearer view of that situation were we to conceive of these problems as a generalized meta-problem or meta-system of problems – a system which I shall henceforth call, the "problématique".[1]

3. Problem relationships

The fragmentation of reality caused by our conceptual (and thus our linguistic) makeup notwithstanding, it is still necessary to talk about the situation and to communicate ideas concerning it. Since we have no new language for doing this, we can only approach the notion of the problematique in terms that are familiar to us.

While keeping in mind that the problematique defines a meta-system of interactive, interdependent, complexifying and proliferating components we are still obliged to illustrate or describe the idea of problematique by naming such components separately. We must still express ourselves in terms of occurrences such as "population explosion", "urban spread, or blight", "illiteracy", "hunger", "environmental dete-

[1] I was forced to make use of this French noun as English does not seem to have a word that corresponds exactly to it. Russell Ackoff has been using the word "mess' to describe the same thing. Although colorful and expressive I don't find this usage a very happy one.

In French, *problématique* refers to the nature of the questions that a philosopher addresses himself to – e.g., "*La problématique kantienne*". However, recently the word has been enlarged as to its connotations, so that in current usage it means sets of ill-structured – or ill-defined – or ill-delimited – problems.

rioration", "energy shortages", "inflation", "pollution", "alienation", "dehumanization", etc.[2]

However, what the notion of problematique aims primarily to convey is not the fact that such problems exist in conjunction with each other, but rather that their interactions create the hard symetries which outline our *predicament* – symetries we seem unable to break. The startling and apparently new fact that inflation, deflation and economic stagnation can co-exist as a composite phenomenon not only defies known remedies but the effects of this combination seem to become more virulent when such remedies are applied to one or the other component. Again, the current, so-called "energy crisis", is connected not only to the geographical distribution of the world's oil resources, but also to the structure of the international monetary system; to the possible reconstitution of a new Palestinian State; to the increasing prices of capital goods; to what various claimants in the conjuncture view to be their self-interest, and to many other similar issues that display confused political, economic, technological, cultural and social dimensions. We know that any partial solutions to this complex happening will inevitably create new clusters of problems – irreversible damage to the environment, to political relations between nations, to current economic structures. That all sorts of current structures should be changed and reconfigured to cope with such dissonances is something to be desired. However, the dynamics of the problematique seems to indicate that it is not change (through adaptation and learning) which is to be expected – it is rather a series of major breakdowns. Therein, of course, lies the danger.

In these two complex occurences, as well as in many others (even simpler ones) the morphology of the problematique can be sensed as a process in which, for whatever reason, problems, be they single or composite, first cluster, then the clusters overlap. Such areas of overlap in turn create new problem-areas, or whole fields, whose description (hence our understanding of them) escapes the boundaries of the original taxonomy. Therefore, the first step in understanding, or rather *penetrating*, the problematique must seek to clarify the systemic configuration of the problem-areas, and secondly, must re-constitute, that is re-design, them in ways that will make their most critical synergies visible.

The five frames on the following two pages attempt a graphic portrayal of the growth of the problematique. In each frame the problem-areas are symbolized by differently shaped shaded spaces. Fig. 1 represents an arbitrary and random positioning of such problem-areas; it describes a situation wherein the visible interactions among the conjuncture of problems are weak or quite probably non-exis-

[2] The original prospectus of the Club of Rome contained a list of forty-nine such components of the problematique, which the Executive Committee had developed, and which I had called Continuous Critical Problems. Today, I find such a list so redundant and static as to be trivial and distortive of the real nature of the problematique, or to be meaningless. Hence I have decided to omit it from the text and to replace it by the above comments, and the several paragraphs that follow them.

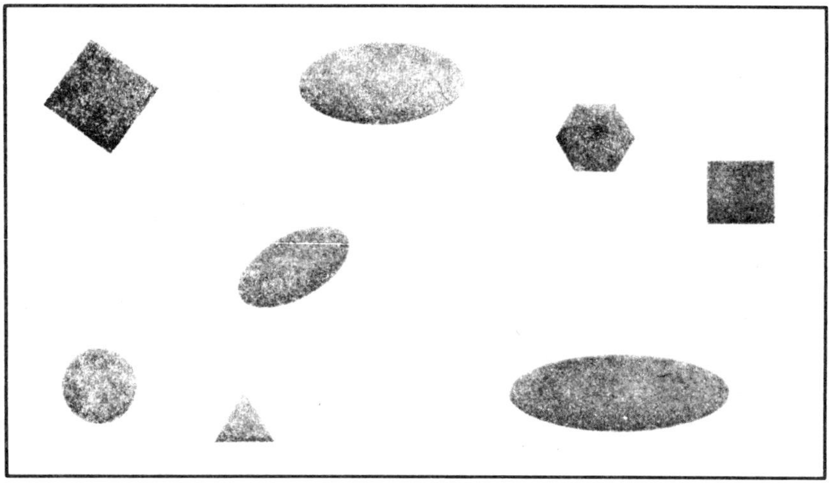

Fig. 1.

tent. In the past such situations were experienced whenever problem-areas were viewed as singular and bounded, albeit occuring in a world context. It is with reference to this type of situation that our traditional analytic problem-solving techniques were developed. But because they were efficient in that type of situation we continue to use them despite our having, by now, recognized the existence of certain contiguities among the problem-areas. Such contiguities are no doubt due to the growth of the problems, or to our deeper understanding of them (fig. 2).

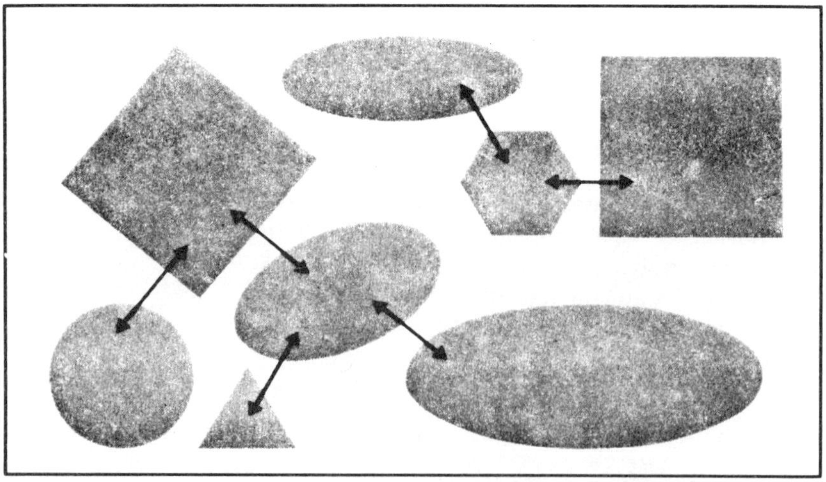

Fig. 2.

Fig. 3.

The continuation of these trends might be said to have turned such contiguities into clusterings and overlaps (fig. 3), which we may perceive superficially but whose real structure and dynamics escape us. In actual fact the situation tends increasingly to appear as a single complex system whose internal relationships, interactions, fields of force, and overlaps are extremely confused (fig. 4) and impossible to delineate without a very serious attempt to model it in its entirety. This kind of modeling effort could, for example, reveal the morphology of the situation

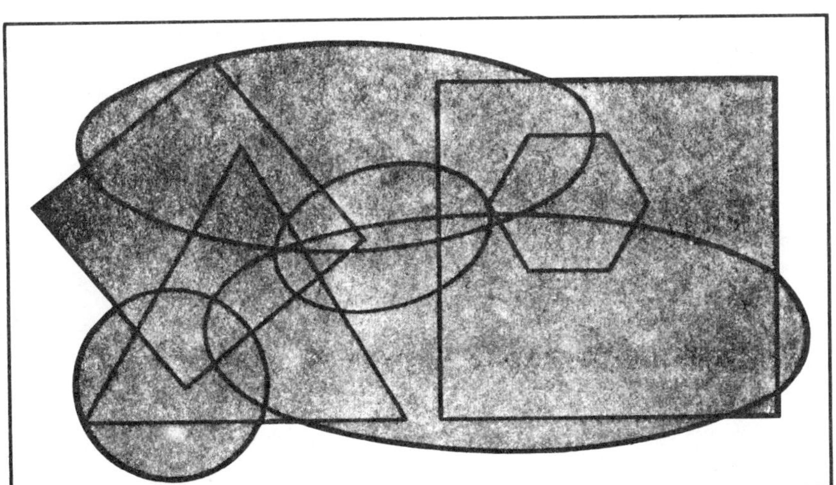

Fig. 4.

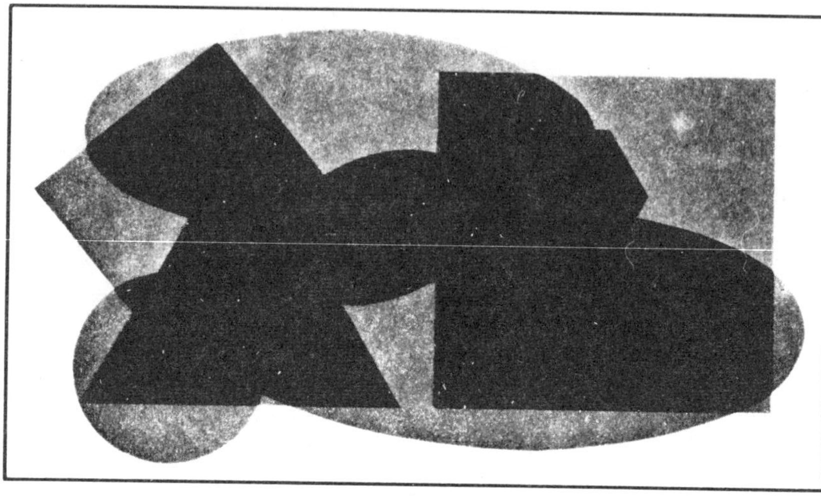

Fig. 5.

as resembling what is shown in fig. 5 – namely, as having a composite dynamic core and differing intensities of interfaces and relationships, all of which must be identified and organized into a unified frame of perception and understanding.[3]

These cursory observations suggest that the causality patterns which feed the dynamics of our situation are extremely complicated and that we have no theory that permits us to identify them *a priori* – namely, before events have congealed into the intractable problematique. And, to attempt to recognize them after the fact does us scarcely any good because, by then, what confronts us bears but scant resemblance to what we have studied, and believe we understood.

If this be true, then, greater effort along customary lines of research should teach us little that is new about the phenomena that make up the issues, and hardly any-

[3] It was after the Club of Rome prospectus was finished that I became acquainted with the works of F.E. Emery and E.L. Trist, especially with their concepts concerning "The general characteristics of social fields". In their discussion of these concepts they distinguish among different typologies of *environment* in a way which appears to be as much richer and more meaningful than what I tried to do in these frames. Yet, there also exists quite a close correspondence between their classification and my description. For instance, fig. 1, in these pages, is explicated in depth by their idea of "Placid random environment"; fig. 2 by their "Placid clustered environment"; fig. 3, by the "Disturbed reactive environment"; and, fig. 5, by what they have named "Turbulent environment". They describe the latter as: "... environments in which there are dynamic processes arising from the field itself which create significant variances for the component systems .... [In these instances] most significance attaches to the case where the dynamic field processes emerge as an unplanned consequence of the actions of the constituent systems ...". F.E. Emery and E.L. Trist, Towards a Social Ecology (Plenum Press, London, 1973).

thing at all about the living, changing dynamic texture of the interconnections that actually create our "situation."

4. Some expectations

At this point we too may have gone as far as we can in general descriptive terms. The questions before us (given the aims and concerns of the Club of Rome) are: What is to be done now? What issues ought to be addressed? How should an effort to address them be organized? What methods of investigation appear pertinent, and what outcomes might one, a priori, hope for?

No direct answers to such questions can be given except to say that the Club of Rome was formed precisely to raise questions and to initiate long-term efforts in the search for "structured responses".

However, because a start must be made somehow, and from some initial point, a general outline of the envisaged effort might be proposed. Thus, the immediate goals of the effort can be stated as follows:

(1) To examine, as systematically as possible, the nature of the profound imbalances that define today's problematique throughout the world, and to attempt to determine the dynamics of the interactions which seemingly exacerbate the situation as a whole.

(2) To develop initial, coarse-grain, models of these interactions in the hope that such models will reveal the systemic components of the situation that are most critical.

(3) To construct a "normative" overview [i.e., theory] from the models and to clarify the action implications – i.e., the political, social, economic, technological, cultural institutional, etc., consequences – that such an overview might entail and substantiate.

(4) To bring whatever that has been learnt as a result of this initial effort, to the attention of those in political authority, in the expectation that such findings might rationally justify consideration of new lines of policy that would be effective in coping with the situation's overall dynamics and its world-wide dimensions.

With respect to these initial goals, it may be argued that many governments and international organizations are beginning to recognize the dangers with which our present situation is fraught; that on the international level bodies such as NATO or OECD are now undertaking detailed work on many individual issues, while the United Nations is planning a world conference on the problems of the "Environment" in 1972. But the point is that the prime difference between these approaches and the one being proposed here resides in the fact that most current efforts continue to be directed toward single or parallel problems and do not attempt any consistent and comprehensive study of the totality of the problematic events that add up to our world system; nor do they address themselves to the areas of dynamic interaction or of overall consequences of these events; nor, for that matter, are they explicitly concerned with questions of institutional change, development, and invention which might be necessary to cope with what is confronting us.

Although one must recognize that modeling is neither new, nor very original, it appears an effective way to trigger the effort. The creation of one or more *situational models* might well be one of the most fruitful ways of learning something new about the world problematique. Such models, in addition to identifying the most critical systemic components of the situation, might reveal – with reference to but almost independently from, the problem-clusters (see fig. 3):

(1) the identity of the most critical and sensitive components of the situation;

(2) the main or major interactions that exist among the various variables contained in the situation;

(3) the behavior of the main variables in relation to changes within the situation;

(4) the temporal ordering of the chief possible outcomes and of their present consequences for action;

(5) the presently invisible critical connections that operate systemically within the present situation and that situation's future configuration;

(6) the positive and negative synergies that must exist among various alternative consequences and options.

Factors such as those listed above can be explored because, by means of modeling the situation, it becomes feasible (1) to penetrate the areas of interdependence among problems and clusters of problems; (2) to manipulate the models artificially, so as to observe the behavior of the situation's components under differently structured configurations.

After the modeling work has been completed it should be possible, perhaps in cooperation with other organizations, to study the critical aggregations revealed by the models with the aim of discovering alternative means of interpreting and resolving interface imbalances and of identifying various options that might serve as guides in the development of coordinated policies.

Such models might also confirm our belief that any desirable, or even acceptable, resolution of the problematique will in all probability entail, at least as outcomes to be seriously considered, fundamental changes in our current social and institutional structures, for the simple reason that these structures were not established to operate in so complex and dynamic a situation as the one in which we find ourselves.

Not a single one of these steps, however, can be taken without establishing a ground upon which the entire modeling effort can be made to rest. Such a ground we shall refer to as the "value-base".

Without such a base we would have no way of determining what situational components are, in fact, critical and what synergies are positive. And we would have no way to talk in terms of corrective or curative action.

5. The value base

The primary aim of modeling is to give a subject a shape, a structure, a configuration that is determined by an end.[4] The clarifications or insights that might be obtained from a successful modeling effort are never reached in terms of the subject (i.e., a problem, a situation or problematique) but in terms of the end to satisfy which the modeling was undertaken in the first place. Such an end always entails a value, and the definition of it must therefore create the particular *value base* that gives meaning and direction to the whole endeavor.

A value-base explicitly stipulates certain assumptions about what is "good" and what is "bad". In the past it was not always necessary to make such a stipulation because a problem could in most instances, be recognized for what it was and therefore fell automatically into a negative value category.

This is not the case nowadays when we must deal with the problematique of a whole world-wide situation. In so extended and complex a problem-field the value premises reveal themselves as being so confused that it becomes imperative to define a value-base that will govern the work from the very outset.

The value-base to be selected must satisfy certain fundamental criteria. First, it must qualify as a tool-concept that can be used throughout the study. Secondly, it must be consonant with the initial perceptions and beliefs that have triggered the effort. Thirdly, it must support, and in some sense justify, the outcomes that are expected from the work. These criteria have already been touched upon; nevertheless, it might bear repeating here that the ground of presuppositions from which we shall start is the belief, backed by considerable empirical evidence, that although the interactions among the events which create our situation are impossible fully to isolate, it should be feasible to identify, through modeling, some critical aspects of their temporal and spatial morphology. And, moreover, that such identification might also permit us to anticipate a number of dissonances which may not exist at present, but whose developing conjuncture could be well forming those new issues which will define our future.

It is on this ground of perceived fact and stated belief that we must now evolve the value-base to work from. This can be done with reference to the attributes of the problematique itself. When we review these (even superficially) we find that all of them are *problems* in *relation* to something else – either to other problem-clusters or to a particular state of the system or to values we take for granted because

[4] Russell Ackoff has described the various meanings of the word "model" as follows: "The word *model* is used as a noun, adjective, and verb, and in each instance it has a slightly different connotation. As a noun "model" is a *representation* in the sense in which an architect constructs a small scale model of a building or a physicist a large-scale model of an atom. As an adjective "model" implies a degree or *perfection* or idealization, as in reference to a model home, a model student, or a model husband. As a verb "to model" means to *demonstrate*, to reveal, to show what a thing is like." (Emphasis in text.) R.L. Ackoff, Scientific Method (John Wiley, New York, 1962).

they are embedded in our current culture. For example, uncontrolled population growth is a problem when viewed in relation to the particular state of the environment that we are now experiencing. It was not a problem when we experienced the environment differently; namely, at a different state of the overall system. Such examples can be multiplied, and if they are we shall notice that in every instance the problematic element derives from an *imbalance* that affects the relationships existing among situational (or systemic) components.

This observation cannot but remind us that imbalance is a state which defines the pathology of an "ecological" system, which, in fact, our situation, seen in its entirety, represents.

Ecology, as one hardly needs to note, is the study of the equilibria and the dynamics of "populations" of living entities within given environments. The notion can be extended and generalized to comprise the equilibria and the dynamics of all entities, for every dimension of contemporary experience is a definable population of facts and concepts: biological, physiological, physical, psychological, ethical, religious, technological, economic, political, national, international, communal, attitudinal, intellectual, institutional, ...; the full list is no doubt finite but very long indeed. It covers every thing and every event among which relations of mutual determination, complementarity or competitiveness can be established.

If we extend, as is increasingly being done nowadays, the definition of ecology to comprise *all* the dimensions of occurence in our environments it becomes possible to say that we are confronted with a problematique which is "ecosystemic" in character. The notion of *balance* is what describes the value-content of any ecosystem. Consequently the idea of *ecological balance* seems to qualify as the appropriate value-base of the study.

Moreover, in our situation, the most generally acceptable and fitting denotation of "the good" is almost self-evidently the re-establishment of the many-dimensional dynamic balance that seems to have been lost in the modern world.

Given the general aims of this study, such a value-base should make it possible to develop models and attain insights that have global relevance, and should further open the way for the integration of the models into a synthesis capable of providing ideas that, subsequently, can be turned into action in terms of policies, of new decision modalities, and of new institutions.

6. Conclusion

To summarize, it can be said that: apart from the reasons of urgency for which the proposed study or project is being recommended, its *a priori* hypotheses arise from the recognition of the problematique as possessing *world-wide* dimensions and therefore *systemic* characteristics, and that the operational attributes of today's world system necessarily involve *normative* elements most of which, being planet-wide, transcend sectoral, political, or regional differences; the recognition that our current methods of explication as well as our social and institutional structures are not designed to operate effectively in a system which is world wide; and that a new and more relevant understanding of our situation –a world view – may lead to more informed and effective policy making on global, regional, national and local levels.

EDUCATIONAL IMPLICATIONS OF RESPONSES TO SYSTEM DYNAMICS MODELS

JAY W. FORRESTER
Massachusetts Institute of Technology

Following the publication of *Urban Dynamics, World Dynamics,* and *The Limits to Growth* several patterns of criticism have emerged that are both repetitive and incorrect. An explanation for the same erroneous conclusions appearing from multiple sources seems to lie in the nature of social science education and the inadequate framework of concepts being created. By briefly examining several published criticisms and identifying underlying issues this paper pleads the need for a stronger educational base in the philosophy of model construction, the nature of dynamic structure in real systems, the use of data in nonlinear feedback models, the objectives of aggregation, the different modes of behavior that can arise as a system shifts from one combination of state variables to another, and the responsibility of authors to maintain relevance of their work to better management of society. This paper does not offer the missing principles, concepts, and philosophy; it suggests some directions for the search.

1. Introduction

Several recent system dynamics models of social behavior have elicited an interesting pattern of unfavorable comment that itself seems in error. Many faulty criticisms appear often enough to suggest that they arise from a widely shared but misleading frame of reference held by the critics. If erroneous reactions are recurring, a possible explanation lies in the failure of our educational processes to deal adequately with the nature of social structure, the processes of formal modeling, and the interpretation of dynamic behavior.

Hundreds of responses followed the appearance of *Urban Dynamics* [8], and thousands, the availability of *World Dynamics* [9] and *The Limits to Growth* [21]. Reaction to the three studies in the public press has tended to be favorable [1, 4, 15–17, 24, 28]. The books have been taken seriously by the public and have already has substantial impact on popular and legislative attitudes toward urban and global problems. But response by the professional and academic press has been split. Some has been very favorable, and many new research projects have been triggered by the three books [3, 12, 13, 19, 23, 26, 31]. But a significant fraction of the academic reaction, especially in the initial book reviews, was negative [5, 6, 14, 25, 29, 30].

All three books were published to offer for debate new methods and propositions about social system structure and behavior. The books all came out early in the research programs from which they evolved. It would not have been surprising if substantial revisions had been necessary to correct for important oversights and shortcomings. But time enough has now elapsed to examine most of the criticisms in considerable depth. The expenditure of much more effort than went into the original writing of the books shows that most of the criticism is itself invalid (see [10, 18, 22, and 27] for some responses).

How could a substantial segment of the criticism repeatedly arrive at the same erroneous conclusions? The reputations of the critics cannot be faulted; the debates have drawn in many acknowledged authorities in their respective fields. How then could such a flood and variety of fallacious commentary be generated? The answer seems to lie in the viewpoint and conceptual paradigm created by conventional social science education. The current educational process apparently does not prepare people for dealing with the behavior of complex, nonlinear, multiple-loop feedback systems. Even worse than leaving a person unprepared, the educational process creates an unjustified self-confidence and a viewpoint that tends to encourage incorrect inferences about dynamic processes and the structures that create change.

This paper takes as examples several of the faulty criticisms and examines their implications about the shortcomings of present social science education. Each issue raised could provide the subject for an independent paper. Space here does not permit a deep analysis of each point made by the critics. Only the direction of rebuttal and explanation will be indicated. For each criticism, I will comment on the apparently faulty underlying foundation. This review does not present the missing principles and concepts that should be woven into social science education, but suggests the need and opportunity for developing such basic understanding of how to create and use models of social behavior.

2. Aggregation in time

In modeling dynamic behavior one can usually aggregate rapid phenomena into structures that have longer-term response, if one is interested only in the slower behavior. By overlooking the relative rapidities of response, many persons find the *World Dynamics* and *The Limits to Growth* models deficient because they omit prices. "The most glaring defect of the Forrester–Meadows models is the absence of any sort of functioning price system" [30]. Free-market prices adjust relatively quickly to clear markets as was seen during 1973 and 1974 in the rapid rise of oil prices. The response time of prices is short compared to the several decades required for changes in population and capital accumulation that are of interest in the two models. When real prices in terms of man-years of effort per unit output are incorporated into a system dynamics model of national economic development [11], the resulting behavior modes are much the same as in the models without prices. There is no dynamically justified reason to believe that full monetary pricing with money flows would substantially alter the long-term modes of the *World Dy-*

namics and *The Limits to Growth* models. But the criteria for aggregating structures that exhibit different response times are not well understood. If the behaviors are close enough in speed of response, they usually entrain one another and become a single response. If they are of sufficiently different response times, they remain separate in their behavior and one can be omitted or coalesced into longer-term behavior if it is not of interest. But even good rules-of-thumb are not available to guide the model maker.

3. Aggregation of similar behavior

When the underlying dynamic structures are similar, it is often permissible to aggregate together system elements that are commonly considered of quite different kinds. In *World Dynamics* [9, p. 53] capital and technology are aggregated together and described, "Capital includes buildings, roads, and factories. It also includes education and the results of scientific research, for the latter are not represented elsewhere in the model system and the investment in them decays at about the same rate as for physical capital." In other words, the additions to capital include not only the physical equipment of production but also education and scientific knowledge. The latter have, to a first approximation, the dynamic characteristics of physical capital. Just as capital is the basis of accumulating more capital, so does knowledge allow the accumulation of more knowledge. Much of scientific and technical knowledge resides in the heads and skills of people and disappears from the system through death in a manner dynamically similar to the obsolescence and discard of physical capital. There is sound logic behind the possibility of aggregating physical and intellectual capital in the interest of model simplicity. But many critics have faulted *World Dynamics* for not including an explicit representation of scientific progress. In rendering such criticism, there has been no argument about the aggregation principles involved or any fundamental development of why the critic believes disaggregation at this point in the system is more important than at the many other points where massive aggregation has been imposed. There are many dimensions to the aggregation issue relative to system dynamics models, but so far, the issues have not been separated and clarified in an orderly way nor have they been placed in perspective with other model-building principles.

4. Aggregation of parallel elements

Whether or not a model-builder aggregates or separates parallel elements in a system structure depends on the purpose of the model. If behavioral characteristics of different elements are essential to addressing the questions to be explored, the elements must be separately represented; if the behavioral characteristics of the aggregate are sufficient to address the question of interest, the separate elements can often be coalesced into a single representation. The subtleties of aggregation have seldom been recognized by the critics of system dynamics models. For example, "one can make the case that the level of aggregation in *World Dynamics* is so high – not distinguishing between developed and underdeveloped nations – that the essential features of the model are lost" [29]. But the essential features of the model are determined by the degree of aggregation; the critic may want a different degree of aggregation and a different purpose, but, if so, he simply wants a different model. In judging a particular model, it must be judged within the purpose for which the aggregation is suitable. In the specific case, the models in *World Dynamics* and *The Limits to Growth* were designed to examine the feasibility of continued growth in global population, capital, and resource usage. These issues depend on the relationships between total world population, total food production, total resource consumption, and total capital, more than on the distribution of population or capital between nations. In general, more adequate guides need to be developed to help the social scientist interrelate purpose and aggregation as a means for judging the particular social models.

5. Past trends versus intrinsic structure

Many types of models are derived from past data and, therefore, can only reflect the trends and modes of the past. In contrast, a system dynamics model is built up from the internal nonlinear structure of the system. The model can be capable of generating modes of behavior that have never been observed in the real system but which may appear at some future time. In differing with the world models, Solow [30] wrote, "As a matter of fact, the relative prices of natural resources and resource products have shown no tendency to rise over the past half-century." Kaysen [14] states, "New ways of locating oil pools and new ways of exploiting them have combined to keep oil reserves – measured in terms of annual consumption – about constant over the past generation, though the actual rate of consumption has been growing exponentially. In general, the relative prices of mineral raw materials and agricultural products have not been rising." Barnett [2] expresses the same, "Forrester's model is unsound in basic structure. He fails to build in feedbacks whereby societies utilize knowledge to overcome resource depletion and pollution .... The real costs per unit of output have declined greatly in the face of population increases. In the U.S. the unit cost of agricultural products, in terms of labor and capital, has fallen by more than half and of mineral products by three-quarters since the Civil War, and the rate of decline has accelerated since World War I." These comments reject a new mode of behavior in the future because that mode has not been observed in the past. Chapters 6 and 8 of *Toward Global Equilibrium* [20] present models of resource utilization and natural gas discovery that show how price is held low and about constant until supply and demand conditions have reached a condition that triggers a precipitous rise in price. The real world seems to be exhibiting the same new mode by breaking out of the past price stability. Education should place less emphasis on past data about behavior and more on the intrinsic structures that can render past trends irrelevant for future management of society.

6. Use of data

System dynamics models have been criticized for failing to incorporate existing statistical data. Yet the techniques commonly used by social scientists to analyze statistical data can be inconclusive or misleading when applied to the kind of non-linear dynamic systems found in real life. One of the issues is illustrated by a recent exchange in the literature. Nordhaus [25] asserts that several assumptions in the demographic sector of *World Dynamics* are contrary to the time-series and cross-sectional data from actual systems. Yet, a rebuttal [10] shows that the model with the allegedly erroneous assumptions generates output data agreeing with the data Nordhaus chose from real life. In other words, the model generates output data that is essentially the same as the data observed from the real world. The data must, therefore, be compatible with the model structure and parameters because it comes from the model as well as from real life. If the data interpretation is incompatible with the parameters of the model, the fault most likely lies in the method of data interpretation. Nordhaus attempts to apply static equilibrium methods to derive from time-series and cross-sectional data the net influence of material standard of living on net population birth rates. However, his method of analysis neglects the simultaneous effect on net birth rates of other variables such as food adequacy, crowding, and pollution. The errors committed by Nordhaus illustrate several typical problems in using statistical data and commonly-applied methods to validate dynamic models.

Although the literature of regression analysis and econometrics dominates the social sciences in describing the use of data for relating real life to models, much new light can be shed on the proper and possible uses of statistical methods by experiments such as Peter Senge is now conducting at MIT (not yet published). In these experiments a system dynamics model is *defined* as the real world, the model entirely endogenously generates time-series data, and standard statistical and econometric methods are used to test whether or not the model-generated data can be used to duplicate the parameter values that are already known in the model. The experiment is completely controllable in terms of the amount of prior knowledge about structure that is available to the analyst, the amount of randomness inherent in the system, the number of variables for which data is available, and the kind and amount of error in the observed data. Preliminary results indicate that an unrealistic degree of perfection is needed in prior knowledge about structure and in the purity of data to keep the data-analysis results from being indecisive or actually misleading. The statistical tests for goodness of fit can be met when, in fact, the computed values are grossly in error and entirely unreasonable. Research and teaching in social science methodology needs to go much more deeply into whether or not the assumptions underlying the conventional methodologies actually prevail in real life.

7. Reverse integration

Running a simulation model backward to see if it retraces past history is often suggested as a way to validate a model. Such was done with the *World Dynamics* model by the Sussex group [6, Chapter 9 and the reply by Meadows *et al.*]. Yet, the procedure is apt to be meaningless unless a subtle combination of conditions falls within a narrow range in which the procedure is informative.[1] Conditions that determine the relevance of backward integration include the purpose of the model, the complexity of its structure, the range of time constants in the structure, how far back in time reversal is attempted compared to the shortest time constants in the model, the presence and magnitude of noise introduced in the model, truncation error, roundoff error, the length of the solution interval, and the nature of the integration algorithm used. In addition, retracing with a model of real-life events is subject to the same limitations that apply to forward prediction [7, Appendix K]. Practical, philosophical, and theoretical considerations overlap in a way that calls for clarification by new theoretical and interpretive work.

8. Author's responsibility

Building models of social systems should be a serious business. The author should believe his own assumptions, procedures, and conclusions. They should represent his best judgment and ability. Furthermore, the author should be expected to present the results of his work as ready for use – either use in social policy or use in some specific sequence of steps toward improved social policy. If the author sees no use behind which he is willing to put his reputation, he should not publish. On the contrary, many articles and criticisms rest on assumptions that the author does not endorse and arrives at conclusions he does not recommend. As an example, Boyd [5] alters the assumptions in the *World Dynamics* model drastically enough to arrive at conclusions different from those in the book. To do so he assumes that an increase by a factor of four in world average technological knowledge will reduce resource usage to zero (absolutely zero) and likewise will reduce pollution generation to zero (likewise absolutely zero). Because Japan and the United States are probably already at or above four times the world average of technological knowledge, Boyd's as-

[1] See the article by Kenneth R. Britting in this volume.

sumption is equivalent to the assertion that those two countries are using zero resources and generating zero pollution. It is reasonably clear that the author does not believe his own assumptions, "The numerical values of these multipliers and constants were invented from whole cloth. They are meant only to reflect, in a reasonable way, the technological optimist's faith that, for example, soon we will be able to find a substitute for any diminishing natural resource . . . . The adherents of this view argue that there are no foreseeable limits on production of goods." But, in fact, the model quickly reverts to the behavior in *World Dynamics* as the assumptions are relaxed slightly from absolutely zero consumption of resources and absolutely zero generation of pollution. This lack of responsibility for how social research couples to the real world appears repeatedly in the professional journals of management science and economics. An author starts from a combination of explicit and implicit assumptions and works through several pages of mathematics built on the assumptions. If one suggests to such an author that the assumptions seem unbelievable in the real-life situation, the author is apt to agree but defends his position simply on the basis that one must start somewhere. He implies that, as long as the assumptions are clear, there is no obligation of responsibility for their applicability to real situations. Until practical relevance is accorded a higher place in judging research in the social sciences, social science will not become more relevant to social practice. Because the standards for judging publication and research originate primarily in academia, criteria of excellence should be moved away from cleverness, mathematical skill, narrow precision, internal logical rigor, and data collection and analysis for their own sakes. Instead relevance to social policy should be expected along with a defense and explanation of that relevance.

9. Structure of lags

Running through the economics literature are repeated assertions that the behavior of a model is critically dependent on the dynamic nature of lags (delays) in the structure of the model. As an example, Solow [30] in discussing the Forrester-Meadows models asserts, "economic quantities are often related to one another with substantial 'distributed lags'. These lag relationships are extraordinarily difficult to estimate from time series, yet the long-run characteristics of the model may be fairly sensitive to the lag structure." However, as has been shown [7, Appendix H], in system dynamics models the structure of delays usually has very little effect on system behavior. These disparate conclusions regarding the sensitivity of models to the formulation of lags seem to arise from basic differences between the econometric and system dynamics approaches to representing social structure and the decision policies that they contain. The educational process should focus on the philosophy of how system structure relates to time-varying behavior and should prepare students to understand the structural concepts underlying different modeling approaches.

10. Rebalancing education and research

Modeling of social systems is being taken more seriously as social stress and complexity increase. Many more people are becoming involved in modeling. Yet, most of the social science literature and education is confined within a few very narrow channels that even together fail to cover the issues or meet the needs of the times. Data is gathered, but for what? Correlations are derived, but to be used where? Models are constructed, but with what commitment to following through to policy implementation? Students learn techniques, but on what foundation of principle and philosophy? Papers are written, but are they for the good of society or merely to display intellectual brilliance for a market that will hire and promote on that criteria? Debates about models proliferate, but with how much sustained follow-up effort to clarify the issues for future guidance of students? There should be more investigation of the underlying principles, concepts, and philosophy of modeling. Such emphasis on fundamentals could well be at the expense of still more data-gathering and model construction until issues of procedure and relevance have become clearer.

References

[1] David C. Anderson, Mr. Forrester's terrible computer, The Wall Street Journal (September 28, 1971).
[2] Harold J. Barnett, World Dynamics, Journal of Economic Literature X, No. 3 (September 1972) 851–854.
[3] Kenneth E. Boulding, The wolf of Rome, Business and Society Review (Summer 1972).
[4] William Bowen, A choice of futures – black or gray, Fortune (September 1971).
[5] Robert Boyd, World Dynamics: A note, Science 177 (August 11, 1972).
[6] H.S.D. Cole et al., Futures (England) 5, Nos. 1 and 2 (February and April 1973), and Models of Doom (Universe Books, New York, 1973).
[7] Jay W. Forrester, Industrial Dynamics (MIT Press, Cambridge, Mass., 1961).
[8] Jay W. Forrester, Urban Dynamics (MIT Press, Cambridge, Mass., 1969).
[9] Jay W. Forrester, World Dynamics (Wright-Allen Press, Cambridge, Mass., 1971).
[10] Jay W. Forrester, Gilbert W. Low and Nathaniel J. Mass, The debate on World Dynamics: A response to Nordhaus, Policy Sciences (June 1974).
[11] Nathan B. Forrester, The Life Cycle of Economic Development, 2nd ed. (Wright-Allen Press, Cambridge, Mass., 1973).
[12] H.E. Hoelscher, World Dynamics, Technos (January–March 1972).
[13] Erich Jantsch, World Dynamics, Futures (June 1971).
[14] Carl Kaysen, The computer that printed out W*O*L*F*, Foreign Affairs 50, No. 4 (July 1972) 660–668.
[15] Richard M. Koff, An end to all this, Playboy (July 1971).
[16] Gerald Leach, Shock findings on the environment crisis, The Observer, London (June 27, 1971).
[17] Anthony Lewis, Nearing the limits: II, The New York Times (October 4, 1973).
[18] Nathaniel J. Mass, Readings in Urban Dynamics, vol. I (Wright-Allen Press, Cambridge, Mass., 1974).

[19] John McLeod, Simulation in the Service of Society, Vol. 1, No. 10, (Oct. 1971) and Vol. 2, No. 3, (March 1972), (Simulation, Box 2228, La Jolla, Calif.).
[20] Dennis L. Meadows and Donella H. Meadows, eds., Toward Global Equilibrium (Wright-Allen Press, Cambridge, Mass., 1973).
[21] Donella H. Meadows, Dennis L. Meadows, Jorgen Randers and William W. Behrens, III, The Limits to Growth (Universe Books for Potomac Associates, 1972).
[22] Meadows et al., Dynamics of Growth in a Finite World (Wright-Allen Press, Cambridge, Mass., 1974).
[23] James M. Moore, "The Limits to Growth" and "World Dynamics", Industrial Engineering (September 1972).
[24] Richard A. Nenneman, Doomsday prophet or man with a message, The Christian Science Monitor (August 7, 1971).
[25] W.D. Nordhaus, World Dynamics: Measurement without data, The Economic Journal, 83, No. 332 (December 1973) 1156–1183.
[26] J.H. Richter, "From economic growth to social progress" and "World Dynamics", World Agriculture XXI, No. 1 (January 1972).
[27] Walter W. Schroeder, III, Robert Sweeney and Louis E. Alfeld, eds., Readings in Urban Dynamics, vol. II (Wright-Allen Press, Cambridge, Mass., 1974).
[28] Harry Schwartz, Forrester's law, The New York Times (June 14, 1971).
[29] S.I. Schwartz and T.C. Foin, A critical review of the social systems models of Jay Forrester, Human Ecology 1, No. 2 (1972) 161–173.
[30] Robert M. Solow, Notes on Doomsday models, Proceedings of National Academy of Sciences U.S.A., 69, No. 12 (December 1972) 3832–3833.
[31] John N. Warfield, World Dynamics, IEEE Transactions on Systems, Man, and Cybernetics (September 1972).

STRUCTURAL GLOBAL MODELS

Frederick BETZ
University of Pennsylvania, USA

and

José Alberto Costacurta de AZEVEDO
University of Pennsylvania, USA, and
Instituto de Pesquisa Espaciais, Brazil

A program for constructing a regionalized global model which is descriptive of the structural detail of the earth is proposed, and the general form of such a model is given in an example. The implications such a program would have for redesigning social institutions are briefly discussed.

1. Introduction

There is an important connection between the ability to use models for global planning and the impact this use would have upon the political and economic structures of our societies. We discuss these questions: (1) what specifications must models for global planning satisfy to be both useful and technically feasible; and (2) what kinds of changes in the social networks of the world will be affected by the development and use of models?

Models can be tested as to truthfulness and accuracy of representation only as they can be used to affect action. In the fundamental methodological sense, it is the consequences that occur to us when we act upon information believed to be true which is the critical test of whether or not the information was true. Global models which allow us to infer the future do so either (1) as a projection of what will occur if all present policies are continued into the future or (2) as a projection of the alteration of some present policy. We must use models to project the future (and not predict the future) because we ourselves are co-producers of our own future.

Thus models for global planning should explicitly display our actions as producer relations in the model. Econometric models fail to do this and their use for planning is thus limited. Leontief and Forrester models do this to some degree but with a commitment to aggregation that limits their usefulness to actual planning decisions or testability of their accuracy in representation. Global models should display the relation of concrete human acts upon the world and upon a regionalized basis so that the impact of actual acts can be studied. Planning consists of designing and implementing desired futures for ourselves through participative procedures. Models must be useful to this process if they are to be scientifically valid.

Currently, many markets operate upon incompleteness of information and differential access to information. Commodity and futures markets are examples of this. If more information about wheat production is available to different wheat producers growing wheat at different times of the year, it will alter their decision to produce (as they expect prices to be affected by their own production). However, valid models of the real world will always affect that world (as the information affects our decisions to act in that world). Any model which has no impact upon the world is not a very accurate or valid model of that world.

2. Technical requirements upon models useful for planning

Models are descriptions and explanations of the behavior of things. They are used to help formulate and attain desired objects. There are three categories of specifications one should place upon planning models: (1) measurability of the descriptive properties of the thing to be modeled, (2) inferential ability of the model to explain the structural interrelations of the modeled thing, and (3) instrumental value of the model in extending a factual base across ranges of different observers with different value orientations. For brevity, we may call these categories: (1) description, (2) explanation, and (3) verification.

2.1. Description

One important use of models for planning is simply to describe accurately and in detail all the complex patterns of interactions that make up the structure of daily life. This is an enormous requirement because data is difficult and costly to collect. The development of satellite-observation techniques has drastically changed this situation. We can now observe relatively cheaply and accurately most physical features of the environment. This means the ecological-environmental systems of the world can be described in detail and the physical resources of economic systems can also be so described. The global models we construct should take these advances into account.

Description in a model differs from explanation in a model, distinguishing between occuring states and relations between occuring states. The general descriptive forms of models are either open-systems or closed-systems. A system is a boundry distinguishing a process from its environment.[1] For an open-system, the states of the environment before and after the effect-of-the-process constitute the descriptive format of the model. These express the existential occurences of the environment that the model summarizes. The process is expressed as an explanatory relation be-

[1] For definitions of systems relevant to problems of models, see Ackoff [4] and Betz and Mitroff [5].

tween these occuring states. For a closed-system, the occuring events are described as the internal states of the process and the relation of explanation between these events summarizes the environment.

Descriptions in a model are summaries of the events which occur in the phenomena which is modeled. When we plan for a system, we wish to know what events are occuring, what events could occur, what event will occur as long as no changes are made in or upon the system (projection) and what events will occur if some change is made in or upon the system (intervention). The structure of phenomena is both the events which occur that is descriptive of the phenomena and the relations between events that is explanatory of the phenomena. This is the structure to be designed, changed, understood, or planned for. Global models should at least describe the occuring structures of society. Since satellite observation makes it possible to describe all physically occuring phenomena, global models should at least provide detailed and complete descriptions of economic-resource, physical, and ecological structures.

This implies that forms of global models be different from the current forms of econometric models, Forrester-Meadows types of models, of Leontief and economic-lot (allocation) forms of models.[2,3] The form of the econometric model is not convenient for displaying the structural detail of economic systems; Forrester–Meadows models do not display the underlying structure of the world system. Both are aggregated indicators of some underlying structure. There is even a serious question as to whether these indicators are the most appropriate summaries of the collections of structures, which can only be answered by determining if these summaries explicitly display which variables control the system and the functional dependence of other variables which complete the description of the system.

Models must help us to control the future (by control, we mean the ability to help us learn of and attain our true desires). If we cannot test the usefulness of these models in controling our own future, then their descriptive completeness and hence their validity cannot be judged. Leontief forms of input–output analysis do summarize structural descriptions on regional levels and can be aggregated to global levels. Accordingly, one can check descriptive completeness on regional levels.

However, in Leontief forms the implicit structure is still not fully made explicit. Each input–output matrix summarizes underlying networks of transformations. These networks are really the structural features one wishes to alter or design in planning efforts. Therefore, a model which displays the networks explicitly and allows one to generate the input–output matrices would be more useful to planning.

[2] Examples of these can be found in Christ [6], Forrester [8], and Leontief [10].

[3] Leontief forms are special cases of allocation problem forms in operations research. These allocation problems are expressed to optimize a decision. What optimality means in a whole world is more complicated a concept than operations research expresses by merely maximizing (or minimizing) an objective function. Accordingly we do not discuss the question of decision rules here, but ask only the question of how to construct an accurate description of some of the significant structures of the world.

Moreover, by not displaying the underlying structural networks, Leontief matrices have committed the modeler to a premature taxonomic scheme. Any general taxonomy used to construct an input–output matrix which is applicable to more than one planning situation will be inefficient for any planning situation. There are always tradeoffs in information between the more general description of the situation and the more specific description of the situation. A model should allow the user to make these tradeoffs in information in the use of the model. Input–output matrices by themselves are inflexible. However, the construction of the underlying networks as graphs in the model and then the generation of the input–output matrices (as adjacency matrices) would provide the needed taxonomic flexibility for planning situations.

2.2. Explanation

A model summarizes the occurences of events that are descriptive of the phenomenum modeled. The summary of the relations between these occurences is the explanatory feature of a model.[4] Two events are said to be correlated if the classes (of which these events are members) exhibit a pattern of related occurences different from a random pattern. A correlation between two classes of events is a directed correlation if and only if the occurence of one class of events is *necessary* to the occurence of the other class of events. The occurence of event A is said to be necessary to the occurence of event B if B occurs *only* when A occurs.

There are two kinds of directed correlations: (1) cause–effect and (2) producer–product. They differ in that cause–effect also accounts for *sufficiency* of directed correlation as an explanation while producer-product *does not* account for sufficiency. Event A is said to be sufficient to the occurence of B if B occurs *whenever* A occurs.

Necessity and sufficiency are the logical criteria which distinguish relations which operate as explanations in our inferences. It is important to distinguish between causal relations and producer relations because models constructed with one or the other relation must be used differently. A causal model can be used to *predict* the future events of a system, whereas a producer model can be used to *prescribe* the future events of a system. Basically, a causal model can be constructed only when the system and its immediate environment is separable from the rest of the universe. In this case, the explanatory relation of cause–effect can not only be necessary but sufficient to account for all the factors in occuring relations – prediction is the explicit detailing of the occurences whose relations are summarized in the causal law; and prediction is possible because nothing else in the universe is required to relate the occurences; and prediction is valid unless the universe intervenes. A producer model does not account for sufficiency. For two events to be producer related to

[4] The concept of explanation in models can be found in Churchman [7] or Ackoff and Emery [3] or Betz and Mitroff [5].

one another, factors in the environment (or universe) of the system must co-produce. In a sense, causal models are contextless models; whereas producer models are contextual, or context-dependent, models. Global models of the world and its futures are dependent upon human acts and therefore dependent upon the contexts of human intelligence and action. Accordingly, global models will be basically producer-models with causal models or quasi-causal models as sub-parts of the model (to the extent some subsystem is separable or quasi-separable).

Producer-relations between occuring events explicitly display the dependence of this relation upon the environment or universe of the events. Human institutions, (political, economic, or cultural) and their design are part of the context of the future of ecology, physical features, or societies of the world's future; so too is science and technology and its development. We must display these dependences explicitly in our models and hence construct producer-models (and not causal-models) of the future. When separability between a system and its universe is possible or conditionally possible, the model so constructed can predict (causal) or project (quasi-causal) the system's future. When separability is not possible, the future occurences can be prescribed to the extent that the co-producing factors can be arranged (producer). Keeping straight the concepts of "prediction", "projection", and "prescription" is essential to avoid misuse or confusion in the nature of the model.

For example, if we model meterological conditions and we do not have the technology to modify weather, then the meterological sphere can be separated from the sphere of human action so that it can be explained causally and hence predicted. As another example, if we model shipping patterns of international trade and *assume* no major policy changes, we can have a quasi-causal model and project the trends in shipping patterns. However, in both cases with appropriate technologies and economic and political institutions, the acts of these institutions can intervene in these predicted or projected futures and co-produce changes in these futures.

Causal and quasi-causal relations express "structural" or "morphological" patterns in the relations of occuring events. Producer relations express "functional" or "purposive" patterns in the relations of occuring events. Both kinds of relations, causal and producer, are available to the global modeler to express phenomenal reality, and the modeler should be conscious of the criteria for their appropriate use.

2.3. Verification

The verification of a model is determined by the degree of control the model has provided in (1) learning about the real and appropriate values of the society and in (2) attaining these values. Any global model which has no impact upon the world has not been tested as to its validity. Better global models can be tested as planning activities become realistic and important. Global models are theoretical tools for social design and governance. As any theory or technique is developed from experience (or experiment or observation of phenomena), so global models ought to be developed from experience in planning situations.

One should avoid a simplistic approach to testing models, in which one makes a prediction from the model and then waits to see whether or not the predictions occur. This is appropriate only for purely causal models; and if the predictions are far into the future, testing may not occur or the testing may occur too late for the model to be of use to anyone. Actions taken in present time, based on information projected (or predicted) by the model, can change or alter those projections (predictions). The use of a global model is to help us design our futures. If we must wait always for the future to test out our model, it will never be currently of any use. If we make no use of it, we will never test its validity.

All models which attempt to infer predictions or projections into the future must be tested in some present sense. We must choose some action on the basis of information inferred from the model and then experience the consequence of that action. If the consequences deviate from what we expected on the basis of the model, we must then analyze the source of the deviation to see if it were from the model itself. If so, then the model must be modified or reconstructed. If there was no deviation, then the model for that present purpose was valid. The broader the range of purposes the model serves instrumentally to infer the consequences of action and the longer over a sequence of actions these purposes are served, the more valid the model becomes in our judgement of how confidently we may believe its inferences.[5] All models are verified or verifiable within the limits of accuracy that the inferred information has been tested. It is therefore essential and fundamental that all models be capable not only of inferring relations between occurences but also expressing the errors on these inferences as a function of the errors of the data feed into the model.[6] This argues that models developed without the cooperation of some agency which acts upon the world can never be verified in any sense that one would believe the information of the model enough to commit one's future upon it.

A current dilemma for the construction of global models is that there exists no global agency which has the ability to affect the whole world at once, such action being a way of directly testing the validity of globally-aggregated models (it may not even be desirable to have such). This emphasizes the importance of regionalized global models that can be tested on a regional basis, so that the global model as a whole will have some reliability as a function of the reliability of the regionalized parts. We then require that global models be testable in actual regionalized planning situations and express the accuracy of their inferrences.

The program of model construction we propose for global models has these requirements:

(1) the models be descriptive of structural details of the world (utilizing the best technology of observation, e.g., satellite techniques);

(2) the models be developed for actual planning situations and thus be at least

[5] Minimal requirements for organizing model testing can be found in Ackoff [2].

[6] A model must be accurate enough to discriminate between the consequence of different policy choices or it is not useful to that decision situation. See Ackoff [1].

as regionalized as there exist governmental authorities or industrial firms to take action upon information which the models provide;

(3) that the regionalized models are connectable into a global model (e.g., organized into a connected computer network);

(4) the models be explicitly constructed in producer or in causal relations and the distinctions maintained in the potential use of information from the model;

(5) the models explicitly express the errors on inferred variables as a function of the errors on observed data.

3. Global structural models

The concept of structure in a model is the relation of parts to each other. A structural relation partially orders the elements of the model set. Topological mathematics is a specialized formal language to express this ordering and the preservation of it as mappings or morphisms. Global models should be topological in form, with the elements of the model indicated as vertices in a network (or graph) and the relations between elements as arcs of the graph.[7] What structural order must be expressed in these graphs so that global planning models can be both regionalized and aggregated to global extent?

The distinction between a regionalized and a global structure is spacial-relation. Transportation flows, which geographers use, are examples of spacial-relation graphs. We should require that global planning models be generalized forms of transportational-flow graphs. For example, meterological conditions can be expressed as spacial movements of air, water, energy, chemical elements, pollutants, etc. In general form, a global model of the world is a vast complex transportational graph wherein elements, energy, people, commodities, things are moved from region to region.[8]

However, to avoid confusion we should note that the ordinary transportation graphs are neither general nor complete enough for our global model. Transportation flows currently do not describe the sources or sinks of commodities transported nor do they trace the different things transported. Geographers have used the concept of a gravity model or of an entropy model (which we *do not* mean to use). An entropy model is a statistically least-biased functional form consistant with the known information about the movement of goods. Gravity models are special cases

[7] Works on graph theory and on transportation flows are many, an example is Potts and Oliver [11].

[8] From causal and producer relations follow two general ways of individualizing nature – in structural or functional classes. Spacial location individuates the structure of physical objects while functional classification relates this structure to human purpose. Regional individuation depends upon purposive classification and connecting these regions together depends upon spacial individuation. These two imageries are discussed in Ackoff and Emery [3].

of entropy models. Both models begin with an aggregated assumption. This is to say, these approaches assume it is too difficult to describe the structural detail of all activities which result in the "attractive" or "repulsive" forces that pull or push goods from place to place. This approach hides the structure for which we should plan and therefore is not useful. We must detail the structure.

How could we detail this structure for better transportational models?

Each vertex in a transportational flow has a structure. Each region in a global transportational flow is represented by a vertex. The structure of the vertex describes the structure of a region. However, as the structure of the global model expresses the spacial-relations (movements of things) between regions, the structure of a region must express relations other than spacial.[9]

Within a region, the relations between elements of that region are producer-product relations. For example, a region is described as a kind of ecological model in which the organisms of the region are classified as producers and consumers; then one graphs the transformations of matter and energy as they pass through networks of living producers and consumers. The composite producer-graph of all organisms is called an ecosystem; and the composite graph of all these producer-relations is called the biosphere. Compiling these eco-system graphs of a region and connecting regions in transportational-flow networks results in a structure which both details the activities of regions and aggregates to global proportions without the loss of any detail. The technologies of satellite observations and computer networks make such a theoretical program practically possible.

The producer-graphs of eco-systems in a region are condensed (by ignoring all cycles in the graph) such that the condensed graph becomes a vertex in the transportational flow connecting regions together (with the degree of that vertex corresponding to all non-cyclic lines in the producer graph). Regions which include human activity require us to generalize the ordinary definition of an ecological-system to include human economic activity. This can be done in a straight-forward way by classifying all industrial and commercial activity as forms of ecological producers and consumers. Any region is then a superposition of graphs of biological eco-systems and economic eco-systems.

Such a global graphical model of the world would have the advantages that (1) it is verifiable in a present sense and as to the degree of accuracy of its representation, (2) it displays the structure of the world which is to be altered or designed if desirable futures are to be brought about, and (3) it is a format which facilitates specific regionalized planning situations within the context of global relations.

The critical taxonomic problem here is the criteria used for regionalizing the graphs. Since there are multiple criteria for regionalization, it is essential that multiple graphs of the same region be constructed. The production at a vertex will be controlled by coproducers that are flows in another superpositioned graph. The

[9] Work in regionalized descriptions has be done by Isard [9].

overall graph for a region or a set of regions (a geography) can be envisioned as a set of superimposed graphs connected to each other in relations of co-production. For an example of a possible graph-set representation of a geography, refer to illustration A. In our work, it has been helpful to distinguish at least three slices: (1) economic, (2) ecological, and (3) physical. We have concentrated upon processes which can be physically individuated. At present, social processes and variables (such as skills, knowledge, laws, customs, etc.) enter the model as exogenous variables. Our feeling is that social processes must be dealt with in models as kinds of logical forms

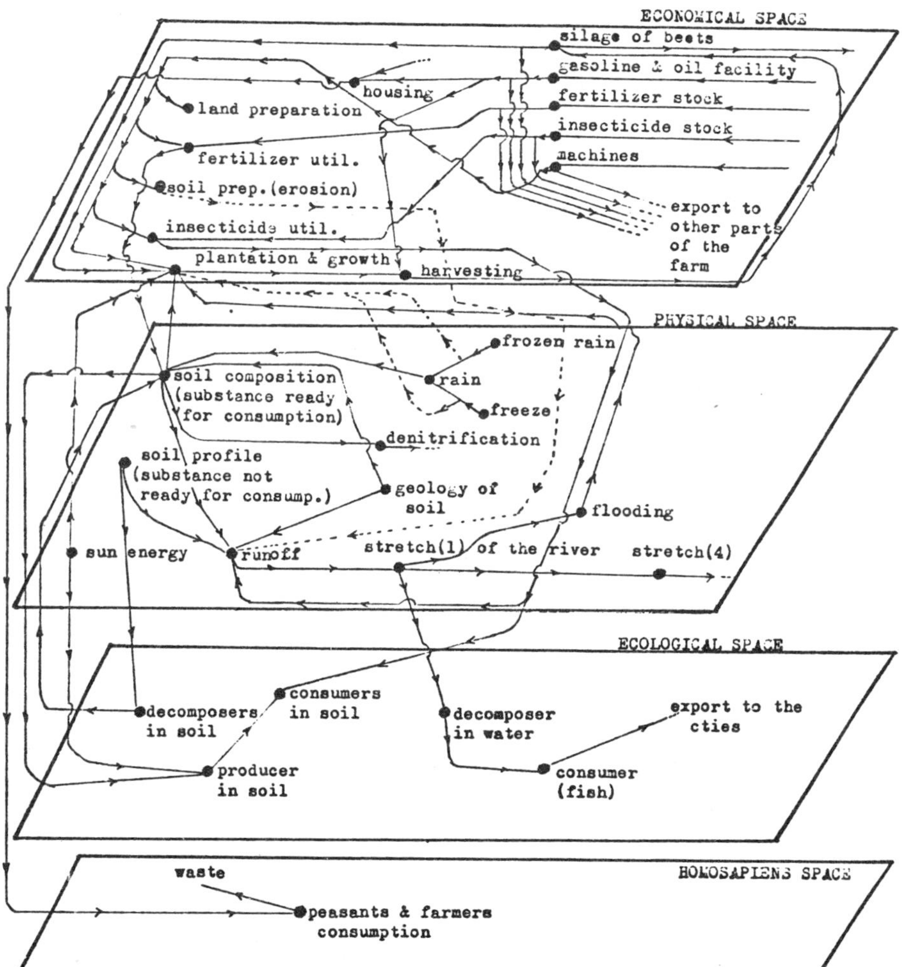

Fig. 1A. Producer graph of region 1 in the picture of fig. 1B.

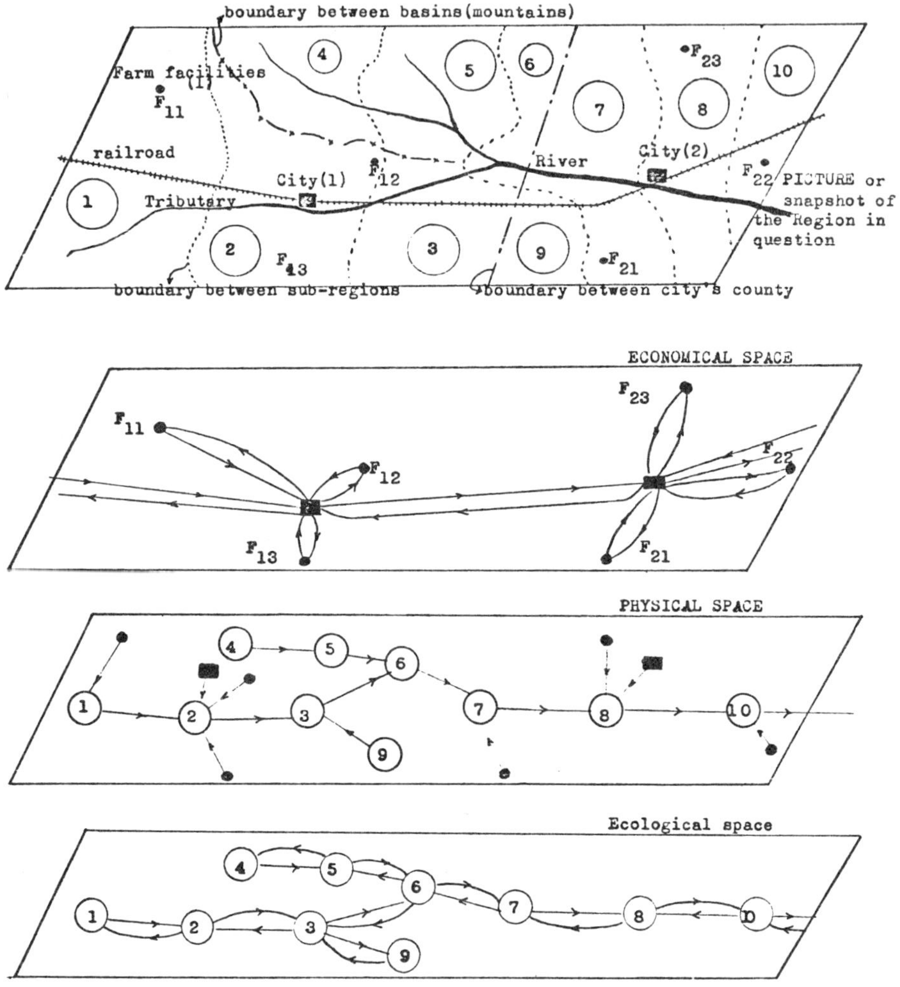

Fig. 1B. Transportational graphs of the geography visualized in the satellite picture.

rather than as merely descriptive processes. We will clarify the way of treating social processes when the design of institutional arrangements to use the model is begun.

4. Impact of global models

Many social institutions currently depend upon the non-existence of information. Differential access to information and uncertain and erroneous information are important factors in various marketing processes. Commodity markets, as an example, allow speculators to gamble on futures prices due to the uncertain quantities that will be produced. Better observations and models will decrease the uncertainty and increase access to information. Those who have vested interests in the current marketing processes are likely to oppose such model development. Such markets should be studied and redesigned to fulfill functions that they are now serving other than gambling or speculation. For example, procedures for noncentral assignment of investment in commodity production must be redesigned.

There will be many institutional-design problems. Governments must organize or reorganize agencies to collect data and model the country on a regionalized basis. An international agency should coordinate these national efforts so that common information can be exchanged between governments. Information will have military, economic, and political implications. Protocols will have to be set up as to what information is exchangable and in what form between nations.

Satellite and computer technology are realities and will not go away, anymore than nuclear weapons would go away no matter how much the physicists who made that technology possible may have wished its disappearance. We must now think about how to design our social institutions so that the impact of global modeling will be beneficial to the world. Differential abilities in modeling should not allow one party or nation to exploit others difficient in that technology. Marketing structures and commercial relations must be regulated, so that information is available and enables such markets to function efficiently. An international agency should take responsibility for the development of regionalized and cooperating networks of modeling and planning efforts. Just as the nuclear stalemate has altered the politics of the world by decreasing the option of world war, let us endeavor to construct an information stalemate which decreases the options of pollution and economic exploitation.

References

[1] Russell L. Ackoff, Scientific Method (Wiley, New York, 1962).
[2] Russell L. Ackoff, A Concept of Corporate Planning (Wiley, New York, 1970).
[3] Russell L. Ackoff and Fred E. Emery, On Purposeful Systems (Aldine, Chicago, 1972).
[4] Russell L. Ackoff, Redesigning the Future (Wiley, New York, 1974).
[5] Frederick Betz and Ian I. Mitroff, Representational systems theory, Management Science 20 (May 1974).
[6] Carl F. Christ, Econometric Models and Methods (Wiley, New York, 1966).
[7] C. West Churchman, Theory of Experimental Inference (MacMillan, 1948).
[8] Jay W. Forrester, World Dynamics (Wright-Allen, Cambridge, Mass., 1971).
[9] Walter Isard, Methods of Regional Analysis (MIT Press, Cambridge, Mass., 1960).
[10] Wassily W. Leontief, The Structure of the American Economy 1919–1939 (Oxford University Press, New York, 1951).
[11] Renfrey B. Potts and Robert M. Oliver, Flows in Transportation Networks (Academic Press, New York, 1972).

PROBLEMS OF DISTRIBUTION

Geoffrey VICKERS
Little Mead, Manor Road, Goring, Reading RG8 9ED, England

1. The Faustian paradox

Men, far more than other creatures, anticipate the future; and because their greater understanding gives them greater power to intervene, they have used this power increasingly to adapt their environment to themselves, rather than to adapt themselves to it. These Promethean, even Faustian powers of understanding and intervening, which once seemed complementary, have proved to be mutually limiting in three main ways.

First, the power to change restricts the power to forecast. Our expectations are based on past experience and are shaken if our actions make the future different. The exhaustion of a familiar resource makes the future less predictable, whether an alternative can be found or not.

Secondly, our power to change the future is shared by others, whose actions are harder to predict or control than the natural course of events. This factor combines with and compounds the first. The shortage and escalating price of oil is due partly to the pressure of escalating demand on existing sources but far more to the political action of middle Eastern states. On a larger scale, this escalation of rival political power has expressed itself in this century in two world wars and countless smaller ones, contained to-day only by a fragile balance of terror.

Thirdly, even the power which any policy maker regards as available to the system which he is helping to manage is itself dependent on the coherence of that system. Human powers of acting and intervening on the scale required to-day involve the co-operation of increasingly large numbers of people over ever longer spans of time. So the power of men to shape the future effectively requires social and political coherence on an ever larger scale. The pursuit of this elusive goal has powered every revolution of our century from Lenin to Castro, from Mussolini to Tito, from Ataturk to Mao-Tse-Tung. Western States, needing more coherence and shared, long-term purpose, are in domestic crisis about how to sustain even what coherence they have.

In consequence, any government and any management which tries to construct a scenario of to-morrow good enough to guide it to-day is bedevilled by the need to anticipate unprecedented change in three dimensions. Even the greatest States, for all their information and all their power, were never so blind or so impotent as to-day, when judged by the standard of understanding, purpose and action which is demanded of to-day by our uncertain and uncontrollable to-morrow.

These sobering thoughts seem to me part of the necessary background to realistic thinking about the nature and scope of forecasting and planning to-day. All forecasts and all plans must take account of all the three elements which I have distinguished.

The first is the field currently explored by studies of *Limits to Growth*. Everyone agrees that current exponential growth rates in the use of fossil fuels and non-regenerative materials, if maintained, will soon exhaust existing sources. Everyone agrees that increasing pollution, like increasing populations, if unchecked, would be self-defeating. Opinions differ fiercely about the extent to which these trends call for restriction and conservation, rather than increased technological development. But either of these and any combination of them will require more or less unfamiliar responses; and the most important, unfamiliar and unwelcome are, I think, the only ones which can be clearly foreseen. To this I will return.

To discuss the impact of these physical changes requires among the disputants some agreement about the dynamics of the world's political, economic and social systems and thus involves the second and third of the elements mentioned above.

2. The limits of international inter-dependence

The surface of the planet, except (at present) the ocean bed, is under the control of some 130 political and fiscal entities. In each of these the inhabitants have access to the resources within its borders and develop and distribute them according to its own political, economic and social system. These systems differ widely; most are highly unstable; and nearly all are in rapid change. They differ also in the problems which each faces. The simplest useful classification of these, I find, is a four-celled matrix, distinguishing along one dimension those which are technologically developed from those which are not and along the other those which are potentially self-supporting at their present population level from those which are not. Britain, for example, though technologically developed, is by this definition in the category of the over-populated.

These distinctions are important because the inhabitants of all these States have far more restricted access to resources outside their borders than to resources within. Their access abroad is dependent on inter-national trade, bilateral barter, development and various kinds of aid and subsidy. In development I include all investment in one country of the resources of others, where this is made for the sake of the economic return to the investor. In aid-and-subsidy I include all other forms of gift or loan, with or without political or other conditions. Development, aid and subsidy appear of course in all manner of combinations.

The sum total of these four devices enlarges to only a limited extent the resources of those States which have little to offer. International trade gives would-be sellers no more claim on external resources than is created in world markets by what they have to offer. Development raises problems for both guest and host country which may prove self-limiting. Aid and subsidy are to-day trivial, when compared with the internal re-distribution of purchasing power which developed countries find necessary to keep even their own economies at work. Estimates of future international relationships must either accept these limitations as permanent or foresee the changes which might transcend them.

Thus although world resources and world production, especially of food, have frightening limitations which are only now coming into view, these are compounded by difficulties of international distribution and it is these distributive difficulties which will shape the resultant problems and which will surely demand unfamiliar responses, economic and political, however well or ill they are handled. On these also the forecaster must take a view. The oil sheiks have usefully shown that these problems will not fall only on undeveloped states.

3. The dynamics of western state systems

Even if Western states were to-day stable systems, neither they nor their business corporations could plan or even prepare for the future without the aid of a fairly comprehensive model of their present internal and external relations; of the dynamics of its development, and of the changes which that development is likely to generate. One critical element in this model would be the dynamics of developing technology. This is sometimes viewed as an infinitely competent and infinitely responsive slave, able and willing to solve any problem presented to it, including those generated by its own growth. But it is also regarded, sometimes by the same people, as a semi-autonomous force, solving problems generated by its own development, and at least as independent in its rate and direction of growth as, say, population.

To resolve this inconsistency would require an analysis of the dynamics of technological growth in the West. This would, I think, reveal three main dynamic foci. The first of these is the pressure on corporate industry to use its accumulated profits in greater and more capital-intensive production. The second is the pressure on Western governments to maintain full employment and a rising standard of living for all. The third is the pressure of public expectation, powered by organized labour, behind all these (I think inconsistent) aspirations. The analysis would also show the hugely increased inter-dependence of government and industry and the newly dominant role of Western governments in financing and directing technological innovation.

Such an analysis would also lead to the main point which I want to make in this paper. I observed earlier that such an analyses would be needed even if Western states to-day were stable systems. But Western States to-day are not stable systems, nor do they combine to form a stable system. On the contrary, they would be faced to-day with the self-generated need for radical change, even if no physical limits to growth were in sight. The cause of this instability, internally as well as internationally, is failure to solve the problems not of production but of distribution.

The hugely productive power of Western economies depends on a most complex distribution of the product, not only between man and man or class and class – those ancient foci of political instability – but also between personal consumption as a whole and the mounting quotient required for investment, public and private and for collective services, two demands which are themselves mutually competitive. In all these fields evidence of failure mounts alarmingly. The following summary is based on British experience but I fear that it applies elsewhere also in greater or less degree.

4. Five fields of failure

Our present ways of settling both the total amount and the differential distribution of personal earnings are absurd. Organised monopolists of labour and of jobs meet to re-negotiate the price of something which neither can do without and neither can get elsewhere; and neither pays for the result of the battle, mis-called a bargain, which is passed on partly to the consumer in increased price and partly to the community in inflation. Thus earnings, though still the most important part of personal income, are neither limited in total nor distributed individually by the rough justice of a free market or by any more refined basis for deciding differential reward. At least nine conflicting ethical criteria are commonly invoked and accepted by both sides in wage negotiations[1] but these serve only to build into the system both instability and inequity, since whatever rewards are won by labour in the capital-intensive sector becomes a valid but unattainable standard for the labour-intensive sector.

To palliate this there has come into being, in America no less than in Europe, a massive system for re-adjusting incomes to redress the grosser inequalities of earnings. This involves recovering through taxation and social contributions and redistributing in cash or kind a substantial slice of GNP, at present perhaps 15–30 per cent, which cannot, I think, fail to grow even larger so long as the inequities and instabilities which it aims to offset are so related as to go on exciting each other as they do now.

Equally inadequate are our present means of distributing GNP between personal consumption, investment and collective services. Collective use and public investment are financed by direct and indirect taxation, by borrowing and by printing money and are decided by acts of political choice made by political executive power within the largely (or wholly) negative control of a legislature. Its largest elements are the oddly assorted couple of defence and education, though increasing claims come from the need to shape the physical environment for the inconsistent purposes

[1] I described these more fully in a paper Changing ethics of distribution, Futures (June 1971) now reproduced as ch. 3 in my book, Making Institutions Work (The Halsted Press, New York, 1974; Associated Business Programmes, London, 1973).

of living in, working in and moving over. The other part of investment, in Britain about one half, is left to the business decisions of national and inter-national firms and is financed increasingly out of profits. Requirements for collective use and public investment are abstracted from the stream of newly generated wealth which would otherwise flow, in profits, wages, salaries, rents or dividends to the sector of personal control and consumption and are thus seen by the elector as a cost.

There results a difference between the standards by which governments and electors measure prosperity. Governments reckon it in GNP but for their electors the critical index to-day is real net personal income. A new car is an asset but a new road is a cost. The dominant importance attached by the electorate to a continuing rise in total net personal incomes imposes a major limitation on the freedom of their governments – and to some extent, of their corporate producers also – to provide for the other two sectors of the national budget, even though this be essential in the longer term to sustain individual incomes, as well as "quality of life". And they are further restricted by the need so to re-distribute part of GNP as to leave the poor richer without making anyone poorer.

All this matters little so long as a country can afford to distribute in total net personal income at least the same proportion of a constantly rising GNP; and this has been the position for a period of some 25 years – not long but long enough to form powerful expectations. It is not the position in the West to-day and it seems to me most unlikely that it will recur for decades, if at all. This radical change seems to me to be the only constant in all the possible scenarios of the future which can rationally be drawn by Western countries to-day.

5. GNP up but incomes down

Those limits to growth which focus so much attention to-day raise problems of distribution more urgent and more important than the problems of limited production; and *they do so whichever disputant proves to be right*. If GNP stabilises or falls, the resultant problems of distribution are obvious. But if the hopes of the expansionists are realised beyond their wildest dreams, the problems of distribution may be just as great. For the expansionist's solution assumes that far more resources will be diverted to investment, including the R & D needed to develop new sources of energy and new materials and to redress pollution by substituting new processes for old. To this must be added the reduced productivity due to working poorer and more scarce resources and the worsening terms of trade between the developed world, as it exhausts its natural resources and the undeveloped world, rich in resources but not yet dependent on them.

Thus even though GNP continues to rise, it seems probable to me that total real net personal incomes will stabilise or fall, not merely for 1974 but for decades and perhaps indefinitely. And this, as it seems to me, is to be expected, whatever mixture of conservationist and expansionist policies be adopted. In any foreseeable conditions any possible rise in GNP will probably be more than fully needed to offset the need to do more work for less return and to meet the other demands which the budget of a developed State cannot avoid in sectors outside personal consumption.

This implies a revolution in personal expectations which will have important implications.

First, all further moves towards greater equality of personal incomes will have to be made by adjustments within that sector. Whatever else growth may still do, it will not enrich the poorer at no cost to the richer. A must take less to give B more. And since in such a siege economy equality will be more strongly demanded, the whole upper half of the income pyramid is bound to be sharply depressed in real terms.

Secondly, the dynamics of growth may be weakened when its product no longer reaches personal pockets, even though growth remains no less important to prevent an even faster fall in personal living standards.

Thirdly, market mechanisms will play an even smaller role than they do now in the complex distribution of resources on which developed economies will still depend; and a correspondingly greater weight will fall on political choice and on the consensus which makes political choice possible.

This will continue a trend already much further advanced than is usually recognized. The mythology of the *free* market still attaches to markets which are ever less free and therefore ever less able to make those "objective" valuations for which markets were once prized. It thus conceals the extent to which markets have already ceased to regulate six of the seven fields of distribution which our grandfathers confidently entrusted to them – goods, services, employment, differential reward, land and capital.[2] The seventh field, foreign exchange, survives only because frantic search has not yet found an alternative. Many names have been invented for the new epoch into which we are peering. No-one, I think, has yet dared to call it the post-market age. Yet this is already its main characteristic, when compared with even a few decades ago. And the price which it exacts is the immense burden of political choice under which we are already labouring. In an age of stable incomes this will escalate still further. Time will show whether the Western genius for political democracy can carry the load.

It is a challenge, alarming, perhaps desperate; but not escapable and surely not ignoble.

[2] I described these more fully in a paper, Whither the mixed economy?, Futures (August 1973).

DEVELOPMENT AND ENVIRONMENT AIMS: AN INTUITIVE VIEW

Jan TINBERGEN
Emeritus professor, Erasmus University Rotterdam, The Netherlands

1. Needed: a synthesis of the Pearson and the Meadows Reports

Two years only after the Pearson Report [1] suggested accelerated growth for the developing world, the Meadows Report [2] told us to decelerate world growth. These two objectives are not necessarily irreconcilable, but very close to incompatibility. Is there a narrow escape? This is the subject of this essay. As we all know, the main phenomena at stake are, for the developed and the developing world, population growth, development of production, with the possible pollution going with it and the exhaustion of natural resources, and the distribution of the fruits of production, both inside and among countries. The danger of incompatibility springs from the strongly felt need for more welfare of the world's poor masses, certainly a billion, soon enough more, on the one hand, and the rapidly increasing dangers connected with pollution of the oceans, rivers and the atmosphere, threatening indispensible food supplies and fresh water and air, on the other hand. Neither of the two forces can be neglected; the only hope is a synthesis of the aims. Quite a few biologists and ecologists doubt the possibility of such a synthesis. An intensive discussion has flared up with as a first result our awareness that we *know too little*.

On *population growth* tendencies we are relatively well informed; we have experienced the consequences of medical improvements without the simultaneous introduction of birth control and shall hardly be able to avoid a doubling of the world population in the early decades of the next century. Quite recently birth control has spread over part of the globe and slightly improved the prospects for Asia and the developed world.

The need for *increased food consumption* in the developing world is too pressing to be denied. The need for higher production in the same countries cannot be denied either; and the unbelievable short-sightedness of the better-to-do all over the world is a major obstacle in this field. It will hardly be possible to deny even to the lower-income groups in the rich countries some improvement of their position. On questions of *distribution of income* within and among countries some useful knowledge has become available in the last five years or so.

Exhaustion of minerals other than fuel does not constitute so much of a threat as was suggested by the Meadows report; on the *energy* prospects opinions widely diverge.

With regard to known types of *pollution* some fairly accurate insight has been gained about the cost of their avoidance and these figures are not too disconcerting either [3]. But ecologists counter with the possibility that new, yet unknown types will be found. By definition there are no figures available on these. The biggest area of uncertainty is the *biosphere*. Some examples in this area of the differences of opinion are the figures produced about the earth's capacity to feed mankind: optimists can be found with FAO [4] and individual experts such as Clark [4], Baade [5], whereas considerable pessimism prevails in biological-ecological circles.

2. The nature of statements to make

With the important lacunae in our present knowledge there must be no misunderstanding about what the nature is of the statements we honestly can make at this stage. Such statements can at best be meant to give an *intuitive picture* of the situation as we see it with the knowledge available and the additional guesses we have made. Any such picture must subsequently be scrutinized with the help of new data fed into a better model. The collection of the data needed will take some years before an appreciable improvement of our knowledge will be reached. The models used so far are remarkable pieces of pioneering work and are in need of improvement from several sides; especially from sciences which have worked with models over a longer period – among them are physics, chemistry and, for the last four decades, econometrics. Many critics of the Meadows report have overlooked the considerable difference between pointing out some scattered errors in some of the equations and calling for the complete revision of a model's results. Some changes in coefficients may have hardly any influence and others may cause a completely new structure.

A second general remark on the nature of statements to be made is similar to a well-known distinction in the theory of socio-economic policy. There is a choice between making *forecasts* and making an *(indicative) plan*. The former assumes no changes in policy and hence becomes utterly unrealistic as soon as periods of more than five years are considered. The latter is difficult enough already and may at least bring out the relevance of a number of aspects of the model. The emphasis is on the possibility only, however.

What I propose to give, then, is my intuitive view as of January 1974. In March I may have had to change my mind because of new evidence that reached me or errors found in my present information or thinking.

3. Rates of growth in production per capita

What I propose to present is a *framework* of some figures of important socio-economic phenomena for the year 2012, showing some consistency and based on an attempt to describe a situation requiring *equal stress in all directions* in which the world socio-economic system is or will be under stress. Of course the year 2012 should not be taken literally; my starting point is the year 1970, for which some relatively reliable information on population and incomes is available [6]. The period of 42 years has been chosen because I attach very high priority to an average annual rate of *5 per cent increase of per capita income* of the developing countries. This means a doubling in fourteen years and a multiplication by eight in 42 years.

The choice of 5 per cent is an attempt to harmonize the opposite pressure between what is needed and what seems possible with the utmost effort. The $3\frac{1}{2}$ to 4 per cent annual growth proposed by the Pearson Report and the official United Nations Development Strategy [7] has been considered too modest by the Columbia Declaration by most of the participants in the colloquium on the Pearson Report, held at Columbia University in February 1970 and organized by Lady Jackson (Barbara Ward). It was rightly pointed out that the rate of 3.5 to 4 per cent per annum would leave many developing countries at an inacceptably low level of living around the year 2000. Having in mind, on the other hand, that the actual performance over the period 1960–1970 in annual rate of growth has been 4.2 per cent for Asia and the Middle East (excl. Japan), 2.2 per cent for Africa (excl. of Libya) and 4.3 for Latin America, a figure of 5 per cent for per capita growth seems to strike the balance between the desirable and the possible.

In view of the impact on resources and pollution the average rate of growth of the *developed* countries of income per capita has been taken at the low level of 1.7 per cent – the figure which doubles that income in the 42 year period chosen. This figure, in my opinion, must be thought of as a figure starting at the present level close to four per cent and declining to *zero quantitative growth* around the end of the period; a figure also higher for the countries with relatively low per capita incomes than for those with the higher incomes. This modest rise is needed to enable the industrialized countries to raise the income levels of their lowest 40 per cent or so. Anyway it has to be accompanied by a further *redistribution of incomes* within countries, where even the socially most advanced among them are able to halve income differences [8]. The instruments needed for such a policy are, on the other hand, better tax policies in the relatively less advanced countries, such as Italy, and, on the other hand, education policies so as to increase the supply of higher qualification and simultaneously reduce the supply of less qualified manpower. These shifts will result in a reduction of income differences and there is some evidence about the extent attainable of this reduction.

4. Rates of growth of population

Again in view of the pressure on natural resources and particularly on the earth's capacity to produce food, a balance must be struck also between this capacity and the size of population around 2012. It seems possible, although again with a considerable effort, to let the population of the developed countries *return to the 1970 figures,* admitting a declining rise to begin with and an absolute reduction later. Recent experiences in almost all developed countries give support to this assumption.

For the developing world such an assumption goes too far. Yet I propose as a goal an average rate of growth, over the period considered, of 1.5 per cent, somewhat more than half the present rate, implying an attainment of a *stable population around the end of the period*, 2012. It needs hardly to be emphasized what efforts of information, education and persuasion will be needed to let this goal materialize. On the other hand the food problem is, in the opinion of most ecologists, of comparable urgency. We will come back to this below.

5. Overall frame of population and income 1970 and 2012

The assumptions set out in sects. 3 and 4 above work out as shown in table 1. The unit of income is the buying power of the 1970 \$. Nominal figures for countries outside the USA have been corrected for the lower price levels by the method suggested by Balassa [9], as far as developed countries are involved, that is, the income per capita difference between the USA and another country has been reduced to $\frac{5}{9}$ of its nominal value. For the developing countries the condition has been added that for a nominal income of zero also a real income of zero must be found. By a crude graphical method an income of nominally \$ 100 becomes \$ 160 in real terms, which looks reasonable.

In order further to characterize the situation implied, the incomes of the *ten per cent richest* and the *ten per cent poorest* regions (defined as groups of countries) have also been estimated and are given in table 2.

These figures are based on the provisional assumption that incomes per capita increase by the same factor 2 in all developed and by the same factor 8 in all developing countries. This is not the ultimate goal, however, and hence the remaining contrasts in 2012 are overestimated.

In order to pass a very crude provisional judgement on the necessity to reduce the

Table 1
Population (in millions), Income per capita (1970 \$ purchasing power) and Total Income (in billions of \$) for developed (*D*) and developing (*U*) countries and totals (*T*), 1970 and 2012.

Year	Population, mln			Income per capita		Total income		
	D	*U*	*T*	*D*	*U*	*D*	*U*	*T*
1970	1089	2588	3677	3700	320	4030	830	4860
2012	1089	4835	5924	7400	2560	8060	12400	20460

Table 2
Incomes (per capita *c* and total *t*) of the richest and poorest decile of countries, 1970 and 2012, and ratios *r* of income per capita of richest to poorest.

	Richest decile		Poorest decile		Ratio [a]
	c	*t*	*c*	*t*	*r*
1970	4400	1620	136	50	32
2012	8200	4870	1060	630	<8

[a] According to Nordhaus and Tobin [11] a better yardstick for the USA per capita income is about $\frac{1}{4}$ lower than GNP per capita. This may imply that the ratio of welfare of rich countries compared to poor is lower by at most $\frac{1}{4}$, or, in table 2, 24 and <6.

ratio of 32 to 1 between per capita incomes in rich and poor countries (the extreme deciles of world population, taken geographically) it is useful to know that inside integrated communities of a stable character, such as the USA, France or Germany the ratio of incomes of the richest to the poorest regions (both covering 10 per cent of the country's population) is about 2 to 1. If we believe, as I do, that by 2012 the world should be somewhat integrated as compared to the present, the ratio of less than 8 to 1 obtained will hardly suffice and may be seen as an indicator of the tense relationships that will remain, even if the "indicative plan" of our figures were carried out – which is doubtful enough.

6. Requirements for agricultural and non-agricultural production

As another illustration let us estimate the necessary agricultural and non-agricultural production for the year 2012. A fair estimate of consumption of agricultural products per capita seems to be \$ 1250 out of an income of \$ 3700 for the rich countries in 1970 and \$ 1750 out of an income of \$ 7400 in 2012. For the poor countries a per capita consumption of 200 out of \$ 320 in 1970 and \$ 850 out of \$ 2560 in 2012 seems reasonable. This leads to table 3.

These figures constitute a 3.2-fold increase in agricultural world production between 1970 and 2012, or slightly less than 3 per cent increase per annum, as com-

Table 3
Per capita (*c*) and total (*t*) Consumption of Agricultural Products of Rich (*D*) and Poor (*U*) countries.

	Dc	*Dt*	*Uc*	*Ut*	Total
1970	1250	1360	200	517	1877
2012	1750	1900	850	4100	6000

pared with the figure of almost 4 per cent in the FAO 1965–1985 World Indicative Plan. As compared with the latter, we have given in to the as yet not quantified anxieties of biologists and ecologists, hoping that they will accelerate their attempts to at least very roughly quantifying their criticism of FAO figures. For all other production – manufacturing, services etc. – we are then left with an increase by almost five times or about 4 per cent per annum, as against the 8 per cent needed for the developing countries according to the DD 2-Strategy.

7. Is zero-quantitative growth impossible?

Among the objections that a number of economists have against the suggestions made by the Meadows Report is that zero-growth in quantitative production would not work. I can't agree with this point of view. If at present we have developed economies with growth rates varying between some 10 per cent in Japan and 3 per cent in the United Kingdom, I don't see why an extrapolation down to zero quantitative growth should be so difficult. In my opinion the large majority of incentives can be maintained in an economy with only qualitative growth [10]. Every society is, as the French say, *un ensemble renouvelé*, meaning that young people have to compete for a job and for replacing the older generation. Also competition among enterprises with a view to improve the quality of their products maintains almost all the incentives.

A zero quantitative growth can be attained only if we shift the incentives downward somewhat by means of government activities such as tax raising and subvention. On later occasions I hope to deal more extensively with the arguments presented against zero quantitative growth in the fall issue 1973 of Daedalus; elsewhere I gave my own arguments in favour of it [10].

References

[1] Lester B. Pearson, Chairman, Partners in Development, Report of the Commission on International Development (New York, 1969).

[2] Donella H. Meadows, Dennis L. Meadows, Jørgen Randers and William Behrens, III, The Limits to Growth (New York, 1972).

[3] For the USA figures have been published in the US News and World Report (an issue before the end of August 1970, not quoted in my source) of $ 71 billion over a period of five years; for The Netherlands a figure of 3 to 4 per cent of national income has been arrived at by Dr. W.J. Beek and this is close to the percentage for the USA and the one estimated for Japan.

[4] Colin Clark, Population (Rome, 1973), quotes FAO planning director Dr. Pawley for having held the opinion that the earth's capacity to grow food could be 50 times as high as around 1970. The figure is not final, I presume (cf. also ref. [5]).

[5] F. Baade, Weltweiter Wohlstand [World Wide Prosperity] (Oldenburg, 1970) p. 158, has come to the conclusion that food production could provide around the year 2000, 6.5 billion people with the "present wasteful level of Western Europe".

[6] World Bank, Trends in Developing Countries (Washington D.C., 1973) (giving also information on the developed countries).
[7] United Nations, International Development Strategy (New York, 1970).
[8] Cf. Jan Tinbergen, Actual, feasible and optimal income inequality in a three-level education model, The Annals of the American Academy of Political and Social Science 409 (1973) 156, where several other studies have also been quoted.
[9] Bela Balassa, Just how misleading are official exchange rate conversions?, Working Papers in Economics, Dept. of Pol. Econ., The John Hopkins University, Baltimore (1973). Balassa corrects an earlier suggestion by David.
[10] Jan Tinbergen, Entwicklungspolitik und Umweltschutz [Development policy and protection of the environment], in: H. von Nussbaum, ed., Die Zukunft des Wachstums (Düsseldorf, 1973) 77. Cf. also my Exhaustion and technological development: A macro-dynamic policy model, to appear in Zeitschrift für Nationalökonomie, Vienna.
[11] W. Nordhaus and J. Tobin, Is Growth Obsolete? Cowles Foundation Discussion Paper No. 319 (New Haven, 1971).

POSSIBLE NEW TAXATION FOR ENVIRONMENTAL IMPROVEMENTS OF THE WORLD

Sumer C. AGGARWAL
The Pennsylvania State University

We know that pollution is caused by production and consumption functions; therefore, it seems logical that producers pay a tax graduated according to the quantity and undesirability of the pollutants they create. In some countries, such taxes are already in existence, and have stimulated the development of antipollution technology; but eventually all the expenditure on pollution control is passed on to the consumer, who is also supposed to benefit from it.

For utilities like power plants (that use fossil fuels or nuclear fuels and that face no competition), spending a reasonable amount on pollution control is a simple matter; because they can pass it on to the consumer without much trouble.

This is not true of other industries. Most producers of the world are under great pressure to offset rising labor and material costs with the help of new technology; but there are no similar pressures for them to avoid polluting the air, water, or land surface. Therefore, to be competitive in the market place, a single company cannot afford to spend any amount on pollution control until all its competitors in the industry do so.

In certain cases, although the origins of pollutants are easily identified, it is not feasible to take direct action against polluters. One example is the millions of car owners who make auto exhaust pervasive. Another is the general public who throw nonreturnable containers on the roadside or on streets. The best way for governments to control such polluters is to tax them at the time they purchase their cars or containers. The taxes so collected may be used to solve the problem by developing new technology to eliminate it or to finance a cleanup agency.

It is not uncommon for masses of people and hosts of small organizations to dispose of many types of residuals and wastes in open environments on a totally random basis. Under such conditions, neither the polluters nor the sources of pollution can be easily identified, because the cost of policing such polluting activities would be prohibitive. The irresponsible actions of these individuals and organizations result in costs of cleaning and social costs to the general public. How can they be made to pay for their irresponsibility?

The elements that degenerate environments are mostly well-known; a few others may become known soon. Therefore our immediate concern is that the scientists and systems analysts be asked to conduct thorough analyses that trace the cause of degenerating elements back from the point at which their pollution effects become visible. These analysts must try to determine which individuals, groups, or organizations are the origins of these problems, and how and why these origins can be held responsible.

If these fail to identify specific origins, then the environmental problems are nobody's individual problems. Because they are harmful to everybody, they are the responsibility of the public in general and of governments in particular. There is a tendency to do nothing until these problems become serious. Then political leaders and thinkers are left with no alternative but to tax the public in general, and to attack the problems just to avoid impending catastrophes.

Meadows et al. [1] are predicting such catastrophes in the future; therefore it may be the appropriate time for us to start thinking about an "*Environmental Improvement Tax*" and to start collecting this tax separately from everybody at a flat rate. If this tax were to be combined with federal taxes, it would leave out many, who contribute as much, if not more pollution than those who do pay. However, it must be recognized that decisions about any new taxation are extremely hard and involve complex political, economic, and social processes and interrelationships.

Economists suggest that under certain special circumstances, some interested parties (who may not even be remotely responsible for generating pollution) may be prepared to pay for antipollution costs, because on an individual basis they can exercize no pollution control, but they will be the major beneficiaries if it is accomplished. For example, property owners near highways may be willing to pay for cleaning the air that deposits dust or corrodes their buildings. Recreationists may be willing to pay on a daily-use basis for checking the pollution of rivers, streams, lakes, and forests. Owners of water-front property may gladly pay for cleaning the adjacent waters. Solid waste disposal may often be paid for by persons who need landfill.

Population explosion is creating much pollution as well as many other social problems. There is no difficulty in identifying those individuals who are origins of exploding population. The main issue is how to tax all these people. In some countries, where population problems have already become severe, incentives are being offered to promote birth control operations or to propagate the use of contraceptive devices; but no penalties have yet been imposed. A logical thing seems to be that the persons who do not take advantage of the services being offered by local governments may be forced to pay taxes for producing more than two children per family. If they are poor or unemployed, they may be made ineligible for welfare and unemployment benefits. Such a proposal brings up a host of humanitarian questions. – Why should the society penalize the newly born? Will not the society be indulging in immoral and irreligious acts? Has the general public any right to regulate an individual's private life, freedom, beliefs, and religion? In some countries, the problem of population has become very severe and gradually a majority of the people are changing their viewpoints. They have now started arguing that if general welfare is the responsibility of the public, then the public has a right to penalize those individuals who do not cooperate in achieving common welfare. If a tax or penalty for producing extra children is ever imposed by some nation, then we can confidently expect a quick decrease in the rate of growth of population of that country.

Fast consumption of nonrenewable resources, and too fast spoilage of arrable lands by multiple cropping each year, with the help of concentrated use of fertilizers, are no longer national problems. These two phenomena have international dimensions and are affecting all the people of the globe. Taxing of agricultural production would have worldwide repercussions; because the poor, who badly need low-cost agricultural products, will be the main victims of an agricultural tax. At present, hardly any thinkers or political leaders seem to support the notion of taxing agriculture. Rather, there is widespread support to agriculture for producing as much as possible just to keep the world population fed.

In a few countries, India, Mexico, Brazil, and some others, during the last two decades land owners have been making large profits; but the small businesses and middle class people have been carrying the major burden of taxation and consequently getting poorer. It is high time that some agricultural tax is imposed on these farmers, particularly on those who are now heavily investing in publicly subsidized agricultural machinery, and dairy-farming machines purely for purposes of huge profits. Such farmers should not be any longer allowed to stay tax-exempt under the disguise of investments for agricultural productivity increases. Agricultural tax as so collected must be pooled with an Environmental Improvement Tax.

To economize in the consumption of nonrenewable resources, taxing of minerals, metals, fossil fuels right at their origins (mines, wells, etc.) seems to be the most rational approach. Whenever and wherever it is economically feasible, efforts towards recycling such substances should be subsidized. Taxes collected from these items may be pooled nationally (or preferably internationally, but that can only be a dream at this stage) for the development of recycling and substitution technologies. A major part of this tax also should go into an Environmental Improvement Tax pool.

Taxes are always repulsive to people. Economists state that too much of taxation leads to decline in disposable income. In turn, the public starts resisting price increases, demand weakens, tax revenues go down, and recession can start. Looking to all such constraints, the only alternative left with the decision-makers is to think of environments as a part of their operations, whether these operations are related to the management of a company, a community, a city, a state, or a nation. They must allocate more and more money collected through Environmental Improvement Taxation (or funds in case of companies) to R and D, for solving environmental problems. The R and D expenses towards improving environments are expected to provide more livable places for the people in general. Their health will improve, there will be less absenteeism, they may be willing to work for lesser wages, overall prices may go down, and finally demand may start strengthening again. In this way, expenditures to improve environments may not only be justified in terms of better quality of life, but also in terms of increased productivity. This author firmly believes that taxes of this nature will encourage desirable attitudinal changes among people and will help them modify their present value-systems in favor of long-term benefits to themselves and their future generations.

Reference

[1] D.H. Meadows, D.L. Meadows, J. Randers and W.W. Behrend, III, The Limits to Growth (Universe Books, New York, 1972).

ON SOME POLICY ISSUES RAISED BY *THE LIMITS OF GROWTH*

Edward S. BOYLAN
Rutgers University

The models underlying both *World Dynamics* [1] and *The Limits to Growth* [2] have been the subject of much study and criticism, perhaps most extensively by Sussex University's Science Policy Research Unit [3]. The authors of [1] and [2] have chosen, for the most part, not to respond to the questions raised by their critics but rather to raise a number of general policy issues. Those criticizing [1] and [2] have rarely discussed these issues, which in some sense could be considered as more "important" than the particular merits and/or shortcomings of the underlying models of [1] and [2]. This, in turn, makes it appear that the viewpoints of Forrester and Meadows are correct.

One of their persistent themes has been that critics have not provided any alternative model. Thus: "It is to be hoped that those who believe they . . . have some different model that is more valid will present it in . . . explicit detail, so that . . . [it] can be examined and compared. To reject this model because of its shortcomings without offering concrete and tangible alternatives would be equivalent to asking that time be stopped" ([1, p. ix], quoted in [4]).

There are a number of reasons why this viewpoint should be rejected. First, it is certainly not true that only those having alternatives are allowed to criticize scientific theories. For example, Linus Pauling does not contend that those who reject Vitamin C as a cold cure must produce their own remedy.

Second, to establish such a criterion would eliminate almost all criticism. The research underlying [2] required the expenditure of millions of dollars. Few individuals (or groups) have the resources for such a project. Moreover, to produce such an alternative would take many months, if not years.

Finally, it is simply not true that critics have not provided alternatives. For in listing areas where the LTG model could be improved they have provided an alternative; to wit: the LTG model incorporating suggested modifications. For example, the inclusion of a feedback loop linking industrial activity with pollution, inclusion of a mechanism for reducing pollution, etc.

Nor is it true that those not offering alternative models are asking that "time be stopped". For the policies advocated in [2] cannot be implemented today in any case. There is no political organization capable of instituting these changes. At best, the proposed policies could be adopted by various nations or groups of nations. Because the LTG model is highly aggregated, there is no guarantee that regional implementation would improve the situation; indeed it might be worse than no change at all. If, for example, economic growth is halted in the developed countries, while population growth continues unabated elsewhere, it is at least arguable that more starvation would occur than if economic growth had continued unabated.

A related contention of Meadows et al. is: "Instead of waiting for perfection models, society must work to construct and implement the best models possible today" [4, p. 218]. The issue, however, is not "perfection". Almost by definition, a model is an imperfect replica of that being modeled. The issue is whether the imperfections are sufficiently significant as to cast doubt on the utility of the model. This writer believes that the imperfections are so great as to make the results of the various simulations quite dubious. Clearly, Meadows and Forrester believe otherwise. This is the crucial issue which should be addressed.

It might be well, at this juncture, to comment on the argument, made especially by Forrester, that public policy hitherto has been based on mental models, which are inherently inferior to computer models at assessing the consequences of proposed (or actual) policies in a dynamic, interrelated system. An obvious example of a dynamic interrelated system is the game of chess. Even only moderately strong human chess players are capable of defeating the best chess-playing computers. The situation could change for more powerful computers with better programming. Clearly, however, there is nothing inherently inferior about a "mental" model of the consequences of a chess move as compared to the explicit computer assessment.

Computer modeling does possess the virtue of forcing the modeler to explicitly formulate his assumptions, relationships, etc. But there is no reason why mental models cannot have similar clarity. Nor should one overestimate the value of an explicit answer. A mental model of "about 6:00 a.m." may be a much better indication of the correct time than a watch stopped at exactly 12:00. Indeed, there is something paradoxical in that the authors of [2] have claimed as one of the prime merits of a computer model that in being explicitly stated it can be more subject to examination and criticism (than a mental model), but at the same time they have rejected almost all criticism on the ground that it was never claimed that the model was "perfect".

But what if the LTG model, warts and all, is the "best" available? Should not public policy then be based on its results? Considering the difficulties in imposing the recommendations flowing from the LTG simulations, it is quite responsible to suggest that strong changes in public policy be made not merely because the "best" model available, a relative term, so indicates, but because a "good" model, an absolute term, so indicates. Only the truly rich can afford to always have the "best". Mankind is not so affluent that it can continually impose new public policies on the basis of the results of every new "best" model.

Such a perspective, to be sure, requires the position that the current status of mankind, if not heaven, is equally removed from hell. Those who view our current status as highly undesirable might well advocate immediate change on the basis of whatever models are now at hand. The authors of [2] clearly view our current status as being very bleak. "Historically, mankind's long record of new inventions has resulted in crowding, deterioration of the environment and greater social inequality because greater productivity has been absorbed by population and capital growth" [2, p. 177]. Many would dispute this assessment. While there is still abject poverty in areas such as India, more people have reached levels of relative affluence than in earlier centuries. In many respects, the ghetto-dwellers of today lead better lives than the kings of medieval Europe.

But what of the future? Are there not "limits to growth" which mankind is ignoring at its peril? It would be foolish to claim that man has infinite resources at his command. But that is not the issue. The question is, are those limits so near as to require immediate changes in public policy?

Clearly, Forrester and Meadows would answer in the affirmative, using the results of [2] as strong evidence supporting this position. Indeed, they often attempt to transform a discussion on the merits of their models into a discussion on whether or not the world is approaching vital limits to growth. But just as the end does not justify the means, the conclusion does not justify the model. When scientists judge models on the basis of the policy positions the model does or does not support, they cease to function as scientists.

Nor is it as clear as Meadows and Forrester would have us believe that the world is rapidly approaching some key barrier to further economic and population growth. Indeed, the true limits may be further away than most might suspect. Let us examine some of the variables which have been viewed as being "limited".

Clearly the stock of natural resources is finite, and eventually there will be reduced supplies of many minerals now abundant. But that day could well be far off. There are large mineral deposits under the seas, and even more minerals are present in dissolved form in sea water. It is now quite expensive, if not impossible, to recover these minerals. If improved technology could allow for economic recovery of these resources, then no shortage of natural resources is likely for centuries. (It should be noted that many of the "disasters" depicted in [2] stem from the assumption that as the stock of natural resources declines, the cost of extracting these resources increases sharply. Thus, it is *assumed* that no new economic means of recovering natural resources will be found.) If such technologies seem exotic and far in the future, one should not underestimate the possibility of new discoveries of natural resources. A major coal deposit was recently discovered in England. And man may learn to substitute more abundant minerals where (now) scarcer metals had been used previously. For example, aluminum is now being used by car makers in some areas where steel was used previously.

There is also great concern regarding continued availability of sources of energy. By some estimates oil supplies will be essentially depleted within fifty years. But, if anything, there is an overabundance of energy sources. Some, to be sure, are quite expensive. There is an enormous difference, however, between "expensive" energy and no energy at all. As one example of an expensive source of energy (under today's technology), it has been suggested that space satellites be orbited which would convert solar power into electrical energy which would then be beamed down to earth using microwaves.

At this point, however, one need not resort to such an expensive scheme. There have been a number of experiments which indicate that in situ processing of oil shale allows for recovering oil at competitive prices while reducing ecological problems presented by other recovery processes. If further, large scale, experiments confirm these findings, we shall have billions of additional barrels of oil available.

The above represent but two large new energy sources which could easily be available in the future. There are a number of other approaches which have been suggested. It does not seem impossible, for example, that a way will be found to use coal as an energy source without producing troublesome amounts of pollution.

Pollution itself could present a threat to economic growth and/or the world's population. But it need not. There is no evidence that even high levels of pollution endanger human lives. (Air pollution does, but this source of "pollution" is coming under control as cars become equipped with pollution control devices and industries install similar equipment.) Moreover, pollution is a relatively new problem and it is far from certain that human ingenuity, i.e., technology, will not prevail against the challenges raised by pollution.

Food is another important variable, one which may present man with his strongest challenge. But the "green revolution" may well be the tip of the iceberg so far as harnessing technology to this problem. New types of grain are being discussed yielding five times (or more) the protein value, per acre, as wheat. In areas such as India yield per acre is only at levels prevailing 30 to 50 years ago in Japan. There is great room for improvement in many regions of the world. An economic means of desalting water could allow millions of acres of land to be irrigated and used as farmland.

Moreover, it is population growth which places the most stress on food supplies. Population growth is much more a problem for the underdeveloped countries than for the West. It is not clear why reduced economic growth by developed countries will increase food supplies or decrease population growth elsewhere. And if population growth were, somehow, reduced in the underdeveloped nations, then food supplies would be much less a source of concern.

One common thread running through these last few paragraphs is the need for new technology to push back limits which might otherwise restrict future economic and population growth. Clearly there is no certainty as to technology succeeding in this task. Meadows and Forrester argue not merely that technology will not succeed, but that technology *cannot* succeed. The goal of these past few paragraphs has not been to argue that technology will succeed, but rather that there are no intrinsic reasons why it cannot be successful.

To a great extent, the history of the past century or two can be viewed as a race between man and technology, with technology fending off the disasters man's actions might otherwise have caused. Nearly 180 years ago Malthus argued that technology must lose that race in the long run. But technology has not lost (at least with regard to food supplies) in the years since Malthus made his arguments. For policy planning purposes, 180 years is certainly the long run. There seem to be no strong reasons to believe that technology *must* now lose that race "in the long run". This is not to say that current policies are perfect and that there are no forms of economic growth which are undesirable. It does say that one need not hastily make sharp changes in public policy out of fear that disaster looms unless changes are made.

In addition to being overly pessimistic as to the course of future events if growth is not halted, the authors of [2] seem to underestimate the shortcomings of life in a "no-growth" society. It is beyond the scope of this paper to examine the less attractive aspects of a no-growth society in detail. It is worthwhile, however, to illustrate briefly how the description of that state in [2] is flawed. We are told there that under no-growth all "human activity that does not require a large flow of irreplaceable resources or produce severe environmental degradation might continue to grow indefinitely. In particular, . . . education, . . . music, . . . basic scientific research . . . could flourish . . ." [2, p. 175]. It would appear, however, that these activities could well require a large flow of natural resources if they were pursued by a *large* number of the population. How are people to be educated, for example, if not through large scale use of books (which would use up large quantities of paper) and expanded construction of school buildings? How are large numbers of people to enjoy music if not through the use of tapes, records, electronic equipment, etc? How is basic research in physics to continue without large scale energy sources?

While technically correct, it is surely misleading to state that "in any equilibrium state the relative levels of capital and population could be adjusted to assure that human material needs are fulfilled at any desired level" [2, p. 175]. The authors provide no discussion as to how the population will be "adjusted" so that material needs are fulfilled at the desired level.

It is surely incorrect to state that among the "technological advances . . . welcome in the equilibrium state . . . [are] medical advances that would decrease the death rate" [2, p. 177]. Once an equilibrium had been reached between the birth rate and the death rate, any new advances would destroy that fragile balance and create considerable social unrest. Far from a blessing, a cure for cancer (say) would be viewed as a grave threat to society at large.

Finally, one should not underestimate that problems which could easily arise in any transition period from our current state to the equilibrium state being advocated. Clearly, there would have to be a redistribution of wealth, for otherwise the poorer countries of the world would not agree to a cessation of world-wide economic growth. Conceivably, the richer countries would stand still until the poorer countries caught up. It is hard to visualize how either process might occur. Indeed, the underdeveloped countries seem blessed primarily with raw natural resources whose value would decrease if economic growth were halted in the developed countries. If they were to attain relative affluence, one would expect considerable demand in those counties for Western inventions such as cars, televisions and air conditioners. It is unclear how this demand could be met or suppressed. As but one minor indication of what troubles might lie ahead, the director of the Social Security system has warned that because of the recent decline in the U.S. birthrate the Social Security system may be running at a sharp deficit (under the current financing set-up) in the 1980s.

References

[1] Jay W. Forrester, World Dynamics (Wright-Allen Press, Cambridge, Mass., 1971).
[2] Donnella and Dennis Meadows et al., The Limits to Growth (Universe Books, New York, 1972).
[3] H.S.D. Cole, Christopher Freeman et al., Models of Doom (Universe Books, New York, 1973).
[4] Donnella and Dennis Meadows et al., A response to Sussex, contained in [3].

INTERDISCIPLINARY MODELING

Olaf HELMER
Graduate School of Business Administration,
University of Southern California

Traditional social modeling, which originally got under way largely in the field of economics and has since spread to many other societal areas, tends to exhibit certain deficiencies which detract from the verisimilitude of the models and hence from their predictive capability. This becomes particularly noticeable when such models are intended to be applied to highly complex systems.

Econometric models of natural economic trends, for example, while of great value in displaying certain relationships between economic variables, generally fail to yield accurate forecasts for the national economic system as a whole. This has been seen very clearly in the attempts to use such models for guidance in the management of the Indian economy; and more recently, in the United States, we have observed the failure to halt inflationary tendencies despite the application of the monetary and fiscal policies suggested by traditional economic models. Similarly, with regard to the system-dynamics approach applied by Forrester and Meadows to the problem of global resources and reported in *Limits to Growth*, there is a widespread belief that even that model, though more sophisticated than traditional ones in some respects, is not fully successful in the manner in which it attempts to represent real-world relationships. The reason for the inadequacy of these models is that they have at least one, and sometimes all, of the following shortcomings:

(i) they are unidisciplinary, rather than multidisciplinary;
(ii) they reflect correlations only, rather than causalities;
(iii) they exhibit trends only, rather than events as well as trends;
(iv) they contain no explicit feedback loops; and
(v) they are deterministic, rather than probabilistic.

These failings, or their positive counterparts (multidisciplinarity, attention to causal effects, inclusion of events, allowance for feedback, and consideration of probabilistic uncertainty), are not entirely independent attributes. For example, an econometric model of a national economy, typically consisting of a set of simultaneous equations (both algebraic and ordinary differential equations), which under "normal" conditions works well and may quite adequately reflect the operation of the real economy, may experience a "system break"[1] caused by an external occurrence that can only be handled within the model by abruptly switching to a different set of input parameters. Such a discontinuity is usually traceable to a non-economic source (see item (i) above), it represents a causal effect rather than a correlation (item (ii)), and more often than not is brought about not by a gradual development (a trend, such as rising expectations) but a point event in time (such as a technological breakthrough, a natural catastrophe, a legislative act) (see item (iii)). Moreover, the reaction of the economic system to such exogenous influences generally has reverse repercussions upon the external system, which, in turn, may produce subsequent shocks to the economy. Yet the model offers no means of reflecting such (positive or negative) feedback (item (iv)); nor can it take into account the uncertainty attaching to exogenous interventions by assigning appropriate probabilities to their occurrence (item (v)).

In order to deal with matters of causality, it is necessary to introduce an explicit time direction and go beyond the simultaneous, merely correlative, interaction of trends. In a model consisting of a set of simultaneous equations, this means that some of these would have to be difference equations or differential difference equations, so as to be capable of relating occurrences at one time to other, "causal", occurrences at an earlier time. This feature, incidentally, is present in the system-dynamics approach, and it permits the introduction of feedback loops, where, say, a rise in trend A causes a subsequent decline in trend B, which in turn may cause a further rise in trend A at a later time. (Thus, feedback loops – item (iv) – presuppose the ability to handle causal effects – item (ii).) The system-dynamics model also is multidisciplinary (item (i)), but is deficient in items (iii) (inclusion of events) and (v) (probability). Of these, the noninclusion of events is likely to be the more significant omission, while the ability to handle probabilistic uncertainty is merely a related, convenient expedient for modeling the occurrence of relevant events.

At present, there does not seem to be a wholly satisfactory method of modeling intricate social systems in such a way as to utilize the powerful traditional simultaneous-equations approach and at the same time include all of the characteristics enumerated here as desirable. However, I do not regard this problem as insurmountable, and in fact there is a new approach being developed that seems to go in the right direction. It offers enough promise to deserve the attention of others interested in modeling complex social interactions, and the purpose of this paper is to present for wider consideration the potential merits of this new approach.

We note, first of all, that the system-dynamics model can easily be extended to accommodate probabilistic features; this means that only one of the five deficiencies mentioned earlier would still apply, namely, the noninclusion of "events", i.e.,

[1] See for example, Henry Winthrop, Social systems and social complexity in relation to interdisciplinary policymaking and planning, Policy Sciences 3, No. 4, pp. 405–420.

occurrences at a certain point in time that cannot conveniently, or even logically, be handled as mere trend fluctuations.

The problem, then, is one of devising a method of modeling the possible occurrence of potential events and of doing so in a manner which permits the adjoining of such an events model, in a natural way, to a traditional trends model of the econometric or system-dynamics type.

The approach to the solution of this problem that I want to present here had its origin in an events-only model developed by Theodore Gordon and myself in connection with a game called "Future", designed on behalf of the Kaiser-Aluminum Corporation in 1967. This approach has since become known as the "cross-impact matrix technique".

The cross-impact idea arose in the context of conducting forecasting studies and reflected an effort to lend a greater degree of sophistication to the estimation of the probabilities of occurrence of potential future events, such as scientific or technological breakthroughs, political events, or acts of legislation. The thought was that a set of potential events, say, $E_1, E_2, \ldots, E_n$, which might be relevant to a given area of inquiry, would not necessarily occur independently of one another and hence ought to be considered in relation to each other. Specifically, the occurrence of one of the events, E_i, might affect the probability of occurrence of another event, E_j. These mutual effects are displayed in a cross-impact matrix, where the cell in which the E_i-row and the E_j-column intersect carries information as to how the probability of occurrence of E_j would be affected if E_i were to take place.

The original cross-impact concept has since been refined and extended in several respects. One refinement consists in giving recognition to the fact that cross impacts occasioned by the occurrence of events must be balanced by cross impacts in the opposite direction in case of nonoccurrence.

Given the probability estimate, p_i, for the occurrence of each event E_i within a given time period, as well as the cross-impact coefficients, c_{ij}, which indicate how much p_j is increased when E_i occurs, it is possible to calculate the effects of nonoccurrence and express them, similarly, as cross-impact coefficients, d_{ij}, where

$$p_i(p_j + c_{ij}) + (1 - p_i)(p_j - d_{ij}) = p_j.$$

Here $p_j + c_{ij}$ and $p_j - d_{ij}$ are the revised probabilities of E_j that result from the oc-

	E_1	E_2	...	E_j	...	E_n
E_1						
E_2						
$\vdots$						
E_i						
$\vdots$						
E_n						

currence or nonoccurrence of E_i, respectively. From this equation it follows at once that

$$d_{ij} = \frac{p_i}{1 - p_i} c_{ij}$$

or

$$d_{ij} = o_i c_{ij},$$

where o_i is the odds for the occurrence of E_i.

The model for the probable occurrence of events $E_1, E_2, \ldots, E_n$, represented by the cross-impact matrix, can now be "run" as follows: Select an E_i at random; decide (by a standard Monte-Carlo procedure) whether it does or does not occur, based on its probability p_i; adjust the probabilities of the other events according to the coefficients in the cross-impact matrix, select another event at random, and repeat the process until all events have been decided. The result will be a "scenario", such as

$$E_3, \bar{E}_1, E_7, E_8, \bar{E}_4, \ldots .$$

Repeating this entire process numerous times (preferably on a computer), the relative frequency with which the event E_i occurs in these scenarios should correspond closely to its prescribed probability, p_i.

By changing the assumed probabilities one at a time and then producing scenarios as before, it is possible to observe the sensitivity with which the outcome depends on the input probabilities. And by changing several input probabilities simultaneously (in accordance with the estimated effect of an assumed action program), it is possible, similarly, to test the effects of alternative action programs. (For example, if E_1 is an act of legislation, the effect of this legislation on the other events under consideration can be examined by setting $p_1 = 1$; or if E_2 is a technological breakthrough, the effect of investing in an R + D effort aimed at bringing E_2 about can be introduced by raising p_2 by an appropriate amount.)

The cross-impact concept has recently been further extended[2] in two important respects: Firstly, cross impacts, which had essentially been handled as correlational effects, have been reinterpreted as properly causal effects. This has been accomplished by subdividing the forecasting period into subintervals and stipulating that the impact of the occurrence or nonoccurrence of an event in one subinterval is not felt until the next subinterval at the earliest. Secondly, trends, in addition to events, are now included in the cross-impact model. Thus, in the extended cross-impact matrix, the entries are "developments", which may be either events or trends.

Here, if the forecasting (or planning) period is broken down into subintervals, or "scenes", $S_1, S_2, \ldots, S_k$ (which may, for instance, be days, or months, or years, or decades, depending on the appropriate and desired temporal "resolution"), estimates

[2] See, for instance: Olaf Helmer, Multi-purpose planning games, Futures 4 (1972).

	$E_1 \dots E_n$	$T_1 \dots T_m$
E_1 ⋮ E_n		
T_1 ⋮ T_m		

will be required of:

(i) the probability of occurrence of each event E_i in each scene S_x,

(ii) the anticipated value of each trend T_j at the beginning of each scene S_x,

(iii) the impact of the occurrence of any event E_i in scene S_x on the probabilities of E_j and the values of T_j in subsequent scenes, and

(iv) the impact of the deviation of any trend T_i from its anticipated value in scene S_x on the probabilities of E_j and the values of T_j in subsequent scenes.

Note that, as implied by the wording of (iv), the place of "occurrence" in the case of an event is taken, in the case of a trend, by a "deviation from its anticipated value".

Given all this information, the model can again be run many items, with or without variations in the inputs required by (i) and (ii), in order to obtain a statistical record of resulting scenarios. The procedure is slightly different from that described earlier because of the causal rather than correlational treatment: In scene S_1, "decide" each of the events; adjust probabilities of all other events in subsequent scenes accordingly (superimposing adjustments by successively applying multiplying factors to the odds of the occurrence of the events), and similarly adjust the anticipated trend values for subsequent scenes (superimposing adjustments by an analogous process); then go on to scene S_2 and proceed similarly, and continue until scene S_k has been played out. Keep a record of event occurrences, by scenes, and of the adjusted trend values. Repeat the entire process sufficiently often to obtain statistically stable results.

Variations in the probability and trend-value inputs will, as before, provide an opportunity for sensitivity studies and for exploring alternative policies and comparing their effects on the resulting scenarios.

To return to the example of Meadows' system-dynamics approach to the study of global resources, this model, as it stands now, is formulated solely in terms of certain trends, primarily the following five: world population, natural resources, food per capita, pollution, and industrial productivity. The rise and fall of these trends, such as a rapid depletion of natural resources, a decline in per-capita food supply, or a dangerous rise in pollution levels, may, in the real world, trigger certain events, such as a breakthrough in opening up new sources of energy (which would be the equivalent of increasing the natural-resources reserves by making the exploitation of lower-grade ores economically feasible), or a breakthrough in the production of non-agricultural food (based on aquiculture or synthetic-protein production, which might avert a serious decline in per-capita food supply), or legislation aimed at ecological control measures (which might reduce pollution to tolerable levels). Yet the Meadows model is incapable of accommodating any such events, except through the device of a system break, that is, the introduction, at a certain point, of new values for some of the parameters occurring in the system's equations.

A first version of a cross-impact model in which potential events such as those just cited are added to the trends incorporated in Meadows' model is currently under construction, and the results of computer runs with this relatively simpleminded model will presently be available. If they look sufficiently promising, this may be an indication that the construction of a more sophisticated cross-impact model of the world resources situation would be a worthwhile undertaking.

Generally speaking, the cross-impact approach, at least in its present form of development, has certain undeniable drawbacks, despite its attractive features. Some of these deficiencies are inherent in its nature, others, it is to be hoped, will eventually be eliminated or at least reduced as the technique undergoes further refinement. A brief examination of some of these shortcomings may be helpful in appraising the potential merits of the method.

First of all, the method is intended to analyze future, rather than past, occurrences. Consequently, it cannot rely on hard, experiential data but must use probability estimates about the future as inputs. These are of necessity soft, judgmental data, which must be obtained, possibly through Delphi-type surveys, from appropriately chosen experts. While some such reliance on expert judgment is unavoidable, this, of course, does not mean that mere intuition is taking the place of sound, theoretical knowledge when the latter is available; indeed, there is nothing to prevent the expert who is asked for his judgment from utilizing whatever theoretical information he may have at his disposal in the form of well-established and well-confirmed models. Thus, even if the mechanism of obtaining the required input data may be highly informal, there is no reason why such data should not be fully representative of the current state of scientific knowledge. In particular, when adjoining events to an existing model of trends and trend interactions (such as the Meadows model), the original, trends-only model may be considered a submodel that loses nothing in rigor by being embedded in the larger, events-and-trends model.

Secondly, there is the matter of the sheer volume of information that is needed. Even a cross-impact analysis of modest size may comprise several dozen developments, requiring thousands of input data in the form of probability estimates, trend forecasts, and cross-impact effects between pairs of developments. To collect such information from a panel of specialists is a sizeable task, and it requires careful organization and possibly the use of computer terminals to obtain well-considered, meaningful entries. In practice, the so-called mini-Delphi procedure, a simplified variant of the regular Delphi technique, has proved quite helpful in this connection.

But the fact remains that the collection of the basic input data for a cross-impact analysis represents a formidable task, though perhaps no more than commensurate with the complexity of the problem that is being tackled.

A serious shortcoming of the method to date is the fact that only the interactions among pairs of developments are examined. In principle, it is possible to extend the procedure to include cases where the occurrence of two or more developments affects the probability of occurrence of another event, but there are two practical impediments to such an extension: the difficulty, even for experts, to state their intuitive perceptions of multidimensional interactions in quantitative terms; and the considerable further enlargement of the volume of the required input data.

Finally, even with regard to merely two-dimensional cross-impacts, there are still some unresolved technical problems. They have to do with: (i) the avoidance of double accounting, which may occur when a development, D_1, has both a direct impact on another development, D_2, and an indirect impact via a third development, D_3; (ii) the balancing, over time, of the impact of the occurrence of an event and the impact of its nonoccurrence; and (iii) the nature of impacts on trends; specifically, whether they should be considered to be causing persistent shifts in their future course, or gradually declining shifts, or merely temporary blips. It may well turn out to be the case that all of these difficulties are due to minor technicalities; yet there is a possibility, which cannot be entirely dismissed, that we are faced with conceptual impediments that may be troublesome to remove.

Despite these shortcomings, of which some are clearly temporary and others possibly inherent, the cross-impact approach, together with the Delphi technique, deserve close examination because they offer a structural remedy to models currently in prevalent use that cannot be attained by the standard methods of refinement usually applied to such models, namely, the disaggregation of model components into subcomponents and the infusion of more precise data. A model can be very effective in mirroring real-world relationships, even if it is relatively coarse and imprecise, provided it includes all the essential features that affect the relationships under consideration. All too often, features are omitted in the modeling process either because it is thought that insufficient data about them are obtainable or because no adequate structural device appears to be readily available for representing them. The realism of an econometric model, for example, may suffer from the omission of vital sociopsychological factors or of technological breakthroughs and acts of legislation. Such omission apparently is made because the inclusion of such items would have required reliance on judgmental, and hence imprecise, information and because there has seemed to be no obvious way in which to adjoin non-economic occurrences to what is basically a purely economic model. In the cross-impact approach there is, by contrast, no barrier to interdisciplinary modeling or to the inclusion of events in addition to the trends customarily treated, and the inclusion of data is judged on the basis of their pertinence rather than their precision.

Not only does it seem that existing models might thus be improved by the introduction of the cross-impact concept, but this approach may even open the way to modeling in areas where researchers have hitherto shied away altogether from the construction of models. Complex social phenomena, such as require careful examination in the context of contemplated legislation, may lend themselves to at least crude cross-impact modeling, by listing relevant events and trends and estimating their interactions. While in many such cases no adequate, wellconfirmed theory may be available that would permit highly reliable forecasts of these developments and their interrelations, the cross-impact technique would compel the researchers to give systematic thought to the pairwise influences among them. And the resultant cross-impact matrix may well represent the next best thing to an as yet nonexistent theory.

CURRENT PERCEPTIONS OF THE WORLD PROBLEMATIQUE

E. FONTELA
Department of Econometrics,
University of Geneva

and

A. GABUS
Battelle, Geneva Research Centre

1. Introduction

Growth in the Western World for the past 20 years has created a growing concern about the future of our society.

A variety of critical problems of considerable complexity have emerged. Among problems sharing a planetary dimension some are of a world nature and require analysis and solutions at the world level; others are widespread and appear as "generalized", common to many areas of the world and requiring locally adjusted solutions; most seem to be somehow interrelated, by multidimensional channels, in a world "problematique" which conditions the predicament of mankind.

A first objective of DEMATEL (*DE*cision *MA*king *T*rial and *E*valuation *L*aboratory, a Battelle Institute, Science and Human Affairs programme, started in 1971) has been to study the changing nature of this problematique.

2. What is a "problem"?

For the purpose of the study, a "problem" has been defined as *a known unacceptable situation, which can be improved by various possible actions.*

As individuals tend to acquire different sets of information about given situations and to have different value systems, the recognition or perception of problems, their intensities and their interrelations, tend to be different for each individual. The study therefore deals with individual perceptions of world problems.

3. Whose perceptions are studied?

To consider that individual actions are a function of problem perceptions seems a sensible working hypothesis. The study has been limited to the analysis of the perceptions of those individuals with a position in society which enables them to synthetise a large amount of information on world issues and to exert influences on actions related to those issues. The number of these individuals is assumed to be small and widely distributed over the world.

4. How are these perceptions explicated?

In order to establish a useful dialogue with influential persons bearing collective responsibility, a "communication procedure" has been designed, tested and applied [1].

This communication procedure has two main components:

(i) a *document* inviting influential persons to communicate their thoughts in response to closed and open questions [2];

(ii) an *interview* based on an analysis of the written answers given under (i).

The *document* contains a suggested list of 48 topics, carefully defined, to which the respondent is invited to react also by rejecting or adding topics. The list of topics was prepared by content analysis of texts dealing with world issues; it is reproduced in Appendix 1.

The questions deal mainly with the intensity of *relations* between the problems accepted by the respondent. Relations are defined as positive impacts of the improvement of a given unacceptable situation on others. Relations are by definition multidimensional (economic, social, political, . . .). The answers of each respondent are analysed in order to derive hierarchical structures, cycles and general organization of the perceived problematique.

For the *interview* that follows the respondent is confronted with the results of the structural analysis of his answers; he can then modify given relations and clarify the nature of some perceived links.

At the end of 1973, 70 persons had received the document; 31 had returned it completed; 25 had been analyzed; 20 interviews had taken place.

5. How to analyze the perceived relations? [3]

In the *document*, the respondents indicate the direct influence that they believe each problem to exert on each of the others according to a scale running by integers from 0 to 4. The answers of each respondent can therefore be arranged in a matrix, X^*, say, of order equal to the number of problems.

This matrix is normalized by multiplying each element of X^* by the reciprocal of the largest row sum of X^*, λ; $(X \equiv \lambda X^*)$. This normalization allows the computation of the matrix of direct and indirect effects $(X(I-X)^{-1})$ and of indirect effects alone $(X^2(I-X)^{-1})$, as well as of vectors of total direct and indirect inferences given $(X(I-X)^{-1}i)$ and received $((I-X')^{-1}X'i)$ by problems.

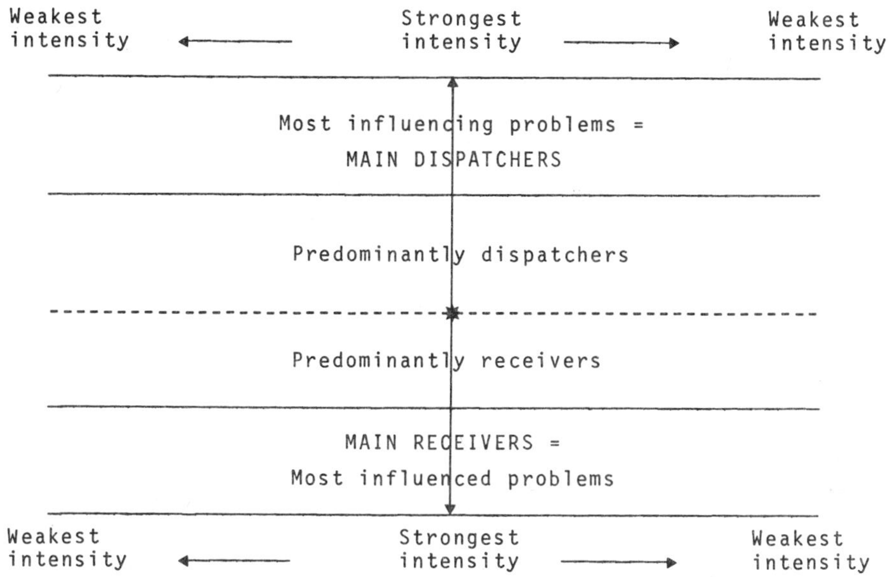

Fig. 1.

This information leads to a "positioning" of problems in a space of intensities given and received, as shown in fig. 1.

It has seemed reasonable to assume that the solution of problems of strong intensity at the bottom correspond to *objectives* of the respondent for improving the situation of mankind, and that problems of strong intensity at the top correspond to *key problems* for which the respondent considers that better answers should be given for attaining the identified objectives (fig. 2).

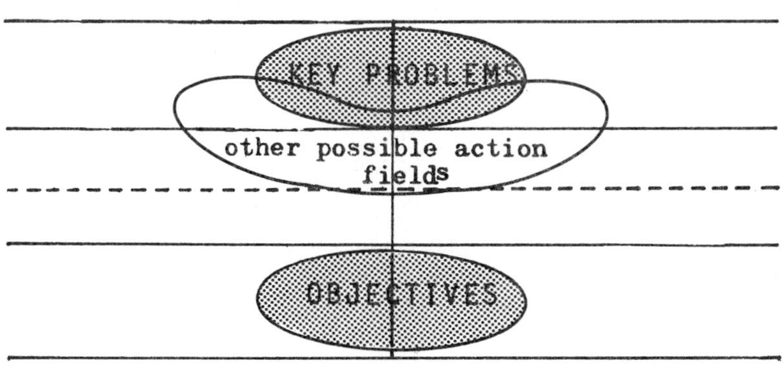

Fig. 2.

The matrix of each respondent has also been analysed in order to derive the geodetic cycles (shortest cycle connecting two problems) and, when possible, the constituents (groups of problems in connected interactive cycles) [4].

A combination of the method of "positioning" with the identification of cycles leads to the drawing of a directed graph of the system perceived by each respondent.

6. Is there a common perception of the world problematique?

The analysis of answers of the first respondents and the interviews that have followed have shown large divergences of views about the perceived structure of the world problematique. It is clear that the perception of key problems and world objectives is highly dependent on the background, position and value system of the respondent. *If* there is a *structure of objectives* behind world issues, it does not seem recognized as such by world bearers of responsibility. The conflictual nature of world society is explained partly by a lack of common "objectives" and partly by the fact that "key problems" are generally viewed in a national perspective, even if their world nature is recognized.

7. What is the minimum basis of agreement?

There is a small subsystem of relations which is accepted by the first 30 respondents as being highly interconnected. This sub-system is shown in Appendix 2.[1] Two constituents are disconnected (their links with the rest of the system are less intense): the problems of less-developed countries and the monetary problems. The other groups of problems (violence, the problems of urban settlements, the preservation of the environment, the depletion of resources, the organisation of world society) are interrelated. It is interesting to note the absence from this graph of problems related to the infringement of the rights of nations (problems 5–6–7) or to the weakening of collective convictions (problem 27) which clearly state the disagreements existing between the value systems of the respondents.

8. Where to go next?

The study of the consensus of the views of all respondents concerning relations between problems is of limited significance.

[1] The distribution of these 30 respondents is the following: North America: 12; Western Europe: 8; Japan: 7; Developing nations: 3.

The first analysis conducted to measure statistical distances between matrices of respondents and to define statistically clusters of respondents has not given significant results [3]. A method or methods of grouping respondents with similar views on goals, key problems, causal chains and cycles needs to be evolved. The aim of this new research effort would be to identify *schools of thought* defined according to types of perceptions of the world problematique.

Appendix 1. Suggested list of world or/and generalized problems

A. Obstacles to world organisation

1. Inadequacy of the institutional organization of international society
2. The difficulty experienced by major powers in defining their interests and responsibilities
3. The arms race
4. Inadequate regulation of use of the sea

B. Infringement of the rights of nations

5. Interference by foreign power
6. Interference by multinational companies
7. Non-completion of decolonization

C. Ineffectiveness of institutions

8. Inability of institutions to adapt to external change
9. Political instability, particularly in the less-developed countries
10. Insufficient personal involvement in political and economic life

D. Difficulties in improving the standard of living of rapidly growing populations

11. Disproportion between the active and inactive population
12. Maladaptation of urban concentrations to individual and collective needs
13. Poor use of available land

E. The crisis of advanced industrial societies

14. Physical and mental illnesses characteristic of advanced industrial civilisation
15. Loss of the sense of personal security
16. Obstacles to fulfilment of non-material aspirations

F. Social discrimination

17. Racial discrimination
18. Recurring threats to the existence of minorities
19. Social discrimination based on sex

G. The use of violence

20. Physical violence in political and social conflict
21. Infringements of fundamental personal freedoms
22. Increase in criminality and delinquency
23. Hijacking and sabotage of aircraft

H. Shortcomings in education and communication

24. Communication difficulties due to complexity and multiplicity of jargon and languages
25. Intensification of political, economic and cultural propaganda in association with the development of mass media
26. Insufficient education of children and adults for an active life

I. The weakening of human motivations

27. Weakening of collective convictions (social, religious, etc.)
28. Difficulties of social advancement
29. Acute disparities between living conditions and aspirations
30. Production, traffic and use of drugs

J. Degradation and disfigurement of the environment

31. Lasting damage to vital properties of water
32. Damage (including noise) to the properties of the atmosphere
33. Damage to vital properties of the soil
34. Disfigurement of the environment: destruction of cultural monuments, natural beauties, etc.
35. Attacks on man's genetic heritage

K. Shortcomings in production and technology

36. Wastage and underemployment of human resources
37. Fundamental waste of material and financial resources
38. Exhaustion of non-renewable mineral and energy reserves
39. Inadequate control of technological development
40. Insufficient efforts to anticipate the exhaustion of energy reserves and to produce less pollutant energy

L. Monetary instability

41. Difficulty in checking inflation
42. Crises in the international monetary system

M. Obstacles to international economic relations

43. Lack of capital for aid to the less-developed countries
44. Low and fluctuating prices of exports from less-developed countries
45. Changes in the flow of international trade due to the establishment of preferential areas

N. Failure to satisfy the basic needs of the less-developed countries

46. Undernutrition and malnutrition in the less-developed countries
47. Endemic and epidemic disease in the less-developed countries
48. Housing shortages and deficiencies in less-developed countries

Appendix 2. Levels and important relations of the directed graph of 30 respondents

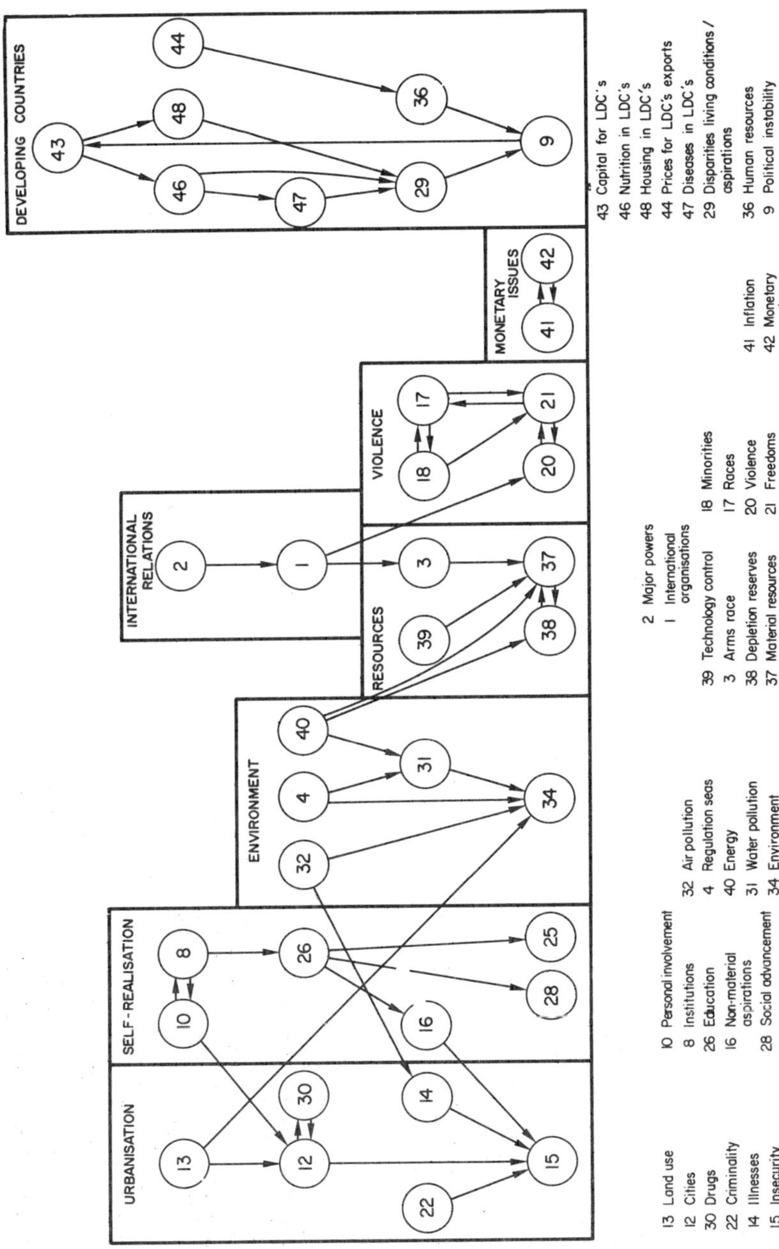

Fig. 3.

References

[1] A. Gabus, Communicating with those bearing collective responsibility, DEMATEL Report No. 1 (Battelle, Geneva Research Centre, 1973).

[2] Battelle, Geneva Research Centre, World Problems – An invitation to further thought (January 1973).

[3] E. Fontela, Analytical methods, DEMATEL Report No. 2 (Battelle, Geneva Research Centre, 1973).

[4] J. Warfield, An assault on complexity, Battelle Monograph No. 3 (Battelle Memorial Institute, Columbus, Ohio, April 1973) and other assays of J. Warfield.

MODELLING THE HUMAN WORLD: PERSPECTIVES

Erich JANTSCH
Center for Research in Management Science,
University of California, Berkeley

Any approach to dynamic systems modelling incorporates assumptions which, in particular, pertain to the nature of the system in question and to the type of inquiry to be represented by the model. A 3 × 3 matrix is proposed for a basic classification of modelling efforts from which emerges also a five-level hierarchy of dynamic systems principles – from causality to creativity – all of which play a role in human systems. The task of modelling the human world then cannot be adequately met by any one approach, but depends on the representation of all levels of dynamic systems principles by designing a *system of approaches* extending over all fields of the matrix.

Dynamic systems modelling which has so far been applied to physical, biological and social systems, but which may also become applicable to cultural (value) and noetic systems, is inherently based on assumptions which are rarely made explicit. The most important of these assumptions pertain to the following questions:

– What is the nature of the system to be modelled? As I shall discuss below, a basic distinction may be made here by using the level of consciousness of the system in question and the ways in which it relates to its environment. In more familiar terms, this leads to the basic distinction between physical, biological/social and spiritual (specifically human) systems.

– What is the position, in relation to the system, from which the model is being constructed? In other words, what is the subject/object relationship, the relationship between observer and observed? Is the system being viewed from the outside, from the angle of view a member of the system might take, or is it viewed as a whole and in the context of its total environment (as a member of a metasystem)? Depending on the view taken, the dynamic aspects of the system appear as change, or process (order of change) or evolution (order of process, or order of order of change).

– What are the parameters in which the outcome is to be obtained? Since they determine the parameters of the whole model, and thus of the inputs to be used, a normative idea is introduced with this choice. What is considered to be maximized or minimized, or kept stationary or at some equilibrium, implies already a certain notion of what is considered as "good" and what as "bad". The higher the level of consciousness, the more important this is, and particularly important with human systems.

Table 1

A matrix of dynamic modelling approaches. Progressing from the lower left (field S1) to the upper right (field A3), a hierarchy of dynamic systems principles emerges, all of which play a role in the human world.

Perspective (level of inquiry) / Nature of system	*Dualistic* Change Control	*Dialectic* Process Regulation	*Evolutionary* Evolution Unfolding wholeness
Apperception Selfreflective consciousness (ternary)	A1 Rational roleplaying models	A2 Learning approaches, gaming, moral systems approach	A3 Formation of spiritual paradigms, cultural morphogenesis, syntony
Perception Reflective consciousness (binary)	P1 Mechanistic system models, rigid behavioristic models	P2 Adaptive system models, homeostasis, ecological models, contingency models	P3 Innovative system models, biological and social morphogenesis, coordinative models of relations
Subjectivity Nonreflective consciousness (singular)	S1 Newtonian mechanics, cause/effect models	S2 Irreversible thermodynamics, statistical mechanics, econometrics	S3 Dissipative structures

Left of the table (vertical arrows spanning the rows): ←Spiritual/human systems→ (all three rows); ←Biological/social systems→ (Perception and Subjectivity rows); ←Physical systems→ (Subjectivity row)

Diagonal labels: Causality (local determinism); Probability (statistically preprogrammed behavior); Vitality (life-preserving behavior); Volition (moral action); Creativity

In the following, I should like to propose a simple characterization of dynamic systems models by using the basic distinctions mentioned under the first two points above, and arranging them in a 3 X 3 matrix (table 1). A look at the nine fields of the matrix, corresponding to nine different approaches to dynamic modelling, permits a clearer understanding of the potentials and limitations of known approaches as well as of yet unsatisfied aspects of the full challenge implied by the ambition of modelling human systems. Each of the nine model categories pertains in some way to the human world – a world which is at the same time physical, biological/social and spiritual, and which we may sometimes regard from the outside; but more often as participating members, and (at least in a numinous context) as members of an evolving human world which is in itself embedded in a greater stream of unfoldment. No single model belonging to any of these categories is capable of adequately representing parts of the human world in which human faculties are played out to the full extent. But by applying different approaches *in the proper perspective*, we may hope to comprehend in a better way micro- and macroaspects of the integral human world.

The *nature of a system* may be characterized by its capability of relating to its total environment (which itself may be composed of physical as well as social and spiritual aspects). Three basic ways, or levels, may be distinguished here:

– A nonreflective, or "blindly unfolding" type of consciousness which may be called *subjectivity*. In a modern systems view [1], all natural systems, from atoms to galaxies, organisms and societies, have subjectivity. The activity of *physical systems* may be fully described at this level.

– A reflective type of consciousness which interacts with the environment through *perception* in a binary system, absorbing experience and forming what Vickers [2] calls a "represented context of reality". Communication is of the signal/response type. The activity of *biological systems* (which are inherently *social systems* at some level) may be described in terms of relations formed in such a binary system of consciousness and environment.

– A selfreflective type of consciousness which reflects also on its own role in the game through a process which, with Leibniz, may be termed *apperception*. It becomes possible by the introduction of a "memory" – or, as Vickers [2] calls it, an "appreciated world" (which, of course, is itself a part of overall consciousness) – in which images, ideas, plans, models and myths may be developed, tested, modified and prepared for realization. Such a ternary scheme forms the basis for an autocatalytic unfolding of consciousness, or *creativity*.[1] *Spiritual*, and in

[1] This seems already recognized in a passage from Lao Tzu's Tao Teh Ching:

> Tao gave birth to One,
> One gave birth to Two,
> Two gave birth to Three,
> Three gave birth to all the myriad things.

(Footnote continued on next page)

particular *human systems* (which are guided by their spiritual dimension manifesting itself in values) may be more fully described only in terms of apperception.

The *views taken of the system*, or *levels of inquiry* represented by a dynamic model, may be characterized by three fundamental modes of the subject/object relationship. I can only sketch them here very briefly[2]:

- In a *dualistic* view, the observer (subject) stands outside the system and perceives *change* expressed through motion of individual system members (objects). He may be compared to a man standing at a river bank, watching the water and what it carries flow by. He measures *quantity*. Interference with the dynamics of the system from the outside is effected by imposing *control*.
- In a *dialectic* view, the observer enters into a feedback relationship with the observed. Out of the manifold interactions among system members emerges an irreversible order of change, or *process*, which also manifests itself in terms of *quality*. Here, the observer may be compared to the captain of a river boat, steering his course in the stream, being its beneficiary as well as its adversary. *Regulation* from inside (the boat) is affected largely by negative feedback action in relation to the environment (the stream), or even by attempting some sort of homeostasis. The dialectic view emphasizes competition, or Darwinian (external) factors of evolution.
- In an *evolutionary* view, the observer becomes one with the observed. He *is* the stream, just as a water molecule or a wave or a whirlpool is part of the stream. Evolution is the unfolding or wholeness through a morphogenetic chain, utilizing energy in manifold ways through positive feedback. Accordingly, the evolutionary view emphasizes what Whyte [5] has called the internal or coordinative factors of evolution. For the human world, such a view assumes a meaningful role of humanity in the context of cosmic evolution – in particular through playing out the human faculties of apperception and conscious design [4]. It cannot be incorporated in a social science conceived in a merely behavioral perspective.

Depending on what aspect of the human world is supposed to be modelled, approaches pertaining to all nine fields of the matrix may be usefully applied, provided they are viewed in the proper perspective. The boundary conditions for their application become even clearer, when we proceed across table 1 from the lower left (field S1) to the upper right (field A3, spanning a variety of *dynamic systems principles* from causality to creativity (finality). We find five levels of such principles:

(Footnote 1 continued)

It may be noted here that, more recently, also Popper's concept of the three worlds of culture-forming man [3] shows some similarity to such a ternary system, but leaves out the feedback relationship between environment and appreciated world. I have further developed the ternary concept elsewhere [4].

[2] I have elaborated elsewhere [4] on the significance of these modes for describing and understanding the dynamics of the human world, as has Prigogine [6] for the physical world.

(1) Field S1: *Causality,* or *local determinism*, which is often effective in describing microaspects of change in a nonsystemic and short-range perspective. Newtonian mechanics belongs here as well as many predictive cause/effect models in the social domain.

(2) Fields S2/P1: *Probability,* or *statistically preprogrammed behavior*, through which may be studied the dynamics of systems for which no potential of selfrenewal is assumed ("linear" forecasting). In physics, field S2 accommodates irreversible thermodynamics and statistical mechanics, whereas in the social domain the bulk of econometric models, based on statistical correlations of past behavior (and implicitly perpetuating such behavior) belongs here. Most of the models applied so far assume equilibrium conditions and define probability accordingly (e.g. in the Second Law of Thermodynamics). The recognition of nonequilibrium conditions in natural systems, for example in nonequilibrium thermodynamics, leads to vastly different concepts and to the morphogenetic principle represented in field S3. The approaches corresponding to field P1 may be termed *mechanistic system models* since no change in system structure is allowed for; in fact, the system is assumed to act in analogy to a Skinner-box, determining the behavior of its members (e.g. no food increase – no population increase). Many types of behavioristic models belong here, from input/output models to most of the published versions of "System Dynamics", including its application to world dynamics [7, 8].

(3) Fields S3/P2/A1: *Vitality*, or *life-preserving behavior*, adaptive to changes in the environment, which has so far been at the focus of General System Theory and has served well to illuminate adaptive, and particularly homeostatic, behavior in the biological and to some extent also in the human social domain.

Field S3 refers to dissipative structures (partially open nonequilibrium systems) which secure their potential for entropy production (energy exchange, metabolism, or life in a general sense) by mutating to ever new dynamic regimes; the underlying principle has been called "order through fluctuation" [9]. What is modelled here – not in a predictive way, since the fluctuations leading to mutations may themselves be random – is evolutionary, morphogenetic dynamics in the physical realm.

Field P2 accommodates approaches to model *adaptive systems*. Ecological models of all kinds, for the nonhuman as well as the human world, belong here, as well as various types of *contingency models* in the social realm. Whereas most models so far take an equilibrium-oriented approach, more realistic nonequilibrium approaches focus on the boundaries and on the conditions for metastability [10, 11]. Contingencies in social models may be introduced either through man-technique interaction, comparing simulation runs of mechanistic system models which use various sets of initial assumptions or constraints (e.g. in the world dynamics studies by Forrester [7] and Meadows et al. [8], or by building a spectrum of contingency responses into the model itself – as has been done for various applications of "System Dynamics" – or by combining the single-level "System Dynamics" approach with a multi-echelon approach "ruling" the responses from a superior coordinative level [12]. In any case, the approach is restricted to a behavioristic outlook, limited by the imagination of the model designer and/or operator.

Field A1 is characterized by rational models of conscious man – the *homo economicus, homo politicus* and other cliches explicitly or implicitly underlying much of current social theory. Here belong the simplifying models of game theory, but also more sophisticated approaches which assume fixed human roles (goals) and adaptive controllable variables in the system, allowing for internal and external fluctuations [13].

(4) Fields P3/A2: *Volition*, which may be considered as the underlying principle of both *moral human action* and (on a much slower timescale) of morphogenesis in biological evolution. It expresses freedom in the focussing of available energy.

Field P3 accommodates models of biological morphogenesis (e.g. the recently published topological approach by Thom [14], which is perhaps the first comprehensive effort in this domain) as well as what may be termed *innovative systems models* in the social realm. No social models of this type have yet been proposed, but the idea would be to formulate something like a principle of "social order through fluctuation" and the conditions for the morphogenetic evolution of systems of relations. There is much to be said for such an order of processes and mutations [4]. Generally, this type of models might be called *coordinative models*, but I would not expect that all dimensions of human relations can be expressed through one model.

A multi-model approach is also essential for field A2 where all kinds of group-learning approaches belong, especially gaming (which, strictly speaking, is not a model, but a "live" simulation experiment). Generally, this is the field for a *moral* or *humanistic systems approach* which attempts to embrace as many views as appear relevant to a systemic situation [15]. An equilibrium approach would seek consensus (often impossible to reach), whereas a nonequilibrium approach would emphasize the recognition of conflict as most essential [16].

(5) Field A3: *Creativity*, or the mutation of human consciousness in morphogenetic ways, generating (or "retrieving" from an unconscious potential) ideas and forms, plans and models, values and motives, works of art and science, physical, social, noetic, ethical and religious systems concepts. It is responsible for man being not only a stabilizer of his world, but also (as John Calhoun puts it) a "crisis-provoker". In the domain of science, ideas for a mutating "order of knowledge through fluctuation" have been proposed by Kuhn [17] and, more recently, by Laszlo [18]. What is badly missing so far, is a useful concept for the mutation of values, or, in more general terms, for the formation of new cultural paradigms and complex systems of such paradigms (religions, ideologies, myths, world views, etc.) – in other words, a model of the "Zeitgeist" – as well as a concept for the mode of communication which leads to a wide sharing of these paradigms and believes ("syntony"). An understanding of the internal (coordinative) factors in the evolution of human consciousness will probably become possible only in the framework of

a wider theory of evolution which recognizes consciousness as a manifestation of an overall unfolding of energy.

These five dynamic systems principles may be understood as forming themselves a hierarchical stratified order, ascending from causality to creativity, from a microscopic mechanism to a macroscopic unfolding touching the numinous; ascending enriches meaning, descending detail (if at the price of reductionism). A tight, predictive model may be constructed only for a world reduced to causality. The farther we climb the hierarchical ladder, the less is one model capable of grasping the fuller dynamics of human systems. Vitality can already only be partially represented through a model to the extent that the model designer or operator is capable of anticipating the spectrum of responses which may come into play. Volition and creativity become even more elusive to modelling and cannot be represented by single models any more. However, they may be approached to some degree through the study of a multitude of models representing causality, probability and vitality aspects in the light of different assumptions, temperaments, descriptions, anticipations, feelings, and insights. Thus, the effort of modelling the human world becomes itself the *design of a dynamic system of approaches* extending over all fields of the matrix in table 1. The more fully we grasp the human world in such a way, the less it is prediction we seek, but an understanding of processes and evolving order in the human world, a recognition of human capabilities as well as a "tuning-in" to evolving order embracing, but going beyond, the human world. Thereby we may hope to improve our capability for *design*, the specific human way of relating to the world.

Most models applied so far to the human world do not go beyond a representation of probability and are geared to the ideal of equilibrium and stabilization. Yet, in a time of transition, such as ours, volition and creativity govern the human world ever more importantly. The task of modelling turns into one of helping the anticipation and realization of social and cultural morphogenesis. The dynamics of the energy unfolding in these realms may be grasped only by recognizing the nonequilibrium nature of human systems and internal coordinative factors in the evolution of relations and human consciousness. There is a wide open field for empirical research as well as theory-building.

References

[1] Ervin Laszlo, Introduction to Systems Philosophy: Toward a New Paradigm of Contemporary Thought (Gordon and Breach and Harper Torchbooks, New York, 1972).
[2] Geoffrey Vickers, Freedom in a Rocking Boat (Basic Books, New York, 1970).
[3] Karl R. Popper, Objective Knowledge (Clarendon Press, Oxford, 1972).
[4] Erich Jantsch, Design for Evolution: Selforganization and Planning in the Life of Human Systems (Braziller, New York, 1975).
[5] Lancelot, L. Whyte, Internal Factors in Evolution (Braziller, New York, 1965).
[6] Ilya Prigogine, Time, irreversibility and structure, in: Jagdish Mehra, ed., The Physicist's Conception of Nature (Reidel, Dordrecht and Boston, 1973).

[7] Jay W. Forrester, World Dynamics (Wright-Allen, Cambridge, Mass., 1971).
[8] Donella H. Meadows et al., The Limits to Growth (Universe Books, New York, 1972).
[9] P. Glansdorff and Ilya Prigogine, Thermodynamic Theory of Structure, Stability, and Fluctuations (Wiley-Interscience, New York, 1971).
[10] Ilya Prigogine et al., Stability, Fluctuations and Complexity, in: Collective Phenomena, vol. 2 (1974).
[11] C.S. Holling, Resilience and stability of ecological systems, in: F. Johnston, Peter W. Frank and Charles D. Michener, eds., Annual Review of Ecology and Systematics, vol. 4 (Annual Reviews, Palo Alto, Calif., 1973).
[12] Mihajlo D. Mesarovic and Eduard C. Pestel, A goal-seeking and regionalized model for analysis of critical world relationships – the conceptual foundation, in: Kybernetes, vol. 1 (1972).
[13] Marshall Moffat, Oral communication on unpublished work related to the design of "adaptive market-adjustment models", in progress in 1973/74 at the Second Institute of Physical Chemistry (Prof. Prigogine), Free University of Brussels. 1974.
[14] René Thom, Stabilité structurelle et morphogenèse (Benjamin, Reading, Mass., 1972).
[15] C. West Churchman, Challenge to Reason (McGraw-Hill, New York 1968), and The Systems Approach (Delacorte, New York, 1968).
[16] Harlan Cleveland, The Future Executive (Harper and Row, New York, 1972).
[17] Thomas S. Kuhn, The Structure of Scientific Revolutions (University of Chicago Press, Chicago, 1962).
[18] Ervin Laszlo, The ideal scientific theory: A thought experiment, in: Philosophy of Science, vol. 40, no. 1 (1973).

A WORLD ISSUE DEBATE: ON ASSUMPTIONS UNDERLYING WORLD MODELS

Richard O. MASON
Graduate School of Management,
University of California

1. The problem

How does one decide on world issues? One method, the dominant one in this collection, is to model the world in some manner and derive the implications of the model. Models, however, have a haunting incompleteness about them. The assumptions which underlie them are seldom explicit. But, more importantly, there is no sure method to guarantee that the assumptions are correct.

In this history of thought the evolution and acceptance of worldviews has probably best been described as a dialogue or a "Great Conversation". One world perspective "has the floor", as it were. It is the reigning paradigm until some other speaker's argument becomes so compelling that the focus of attention is shifted to it and a new world perspective takes charge.

When a modeler considers the problem of constructing world models he is faced with the problem of choosing the worldview which it presumes. This, I believe, is best accomplished by means of debate on the underlying assumptions of the model. But not just any debate. As scientists we should seek to develop an optimal debate, one that is more efficient and effective than the loosely knit dialogue of centurys. We must compress the time involved in "Great Conversations" and sharpen the controversy to highlight better the relevant points of concern.

What follows is an attempt to construct such a worldview debate. I lay no especial claim to its efficiency or effectiveness. It is too early in the work concerning the design of rational debate to make such claims. However, I do submit that this debate, or one like it, must be resolved before an appropriate world model can be constructed. All world models must deal, implicitly at least, with these issues.

This sample debate focuses on land-use policy in Southern Los Angeles County from the period 1940 to 1970, and as projected to 1990. It is an archetype, however, for a debate that any world modeler might use. The data for the debate are contained in table 1.

Table 1
Land use trends in Southern Los Angeles County[a]

Land use categories[b]	Past 1940	Present 1970	Forecast 1990
Urban used	25.1%	73.6%	82.0%
Urban vacant	10.1%	5.3%	3.2%
Nonurban used	45.3%	7.3%	4.3%
Nonurban vacant	19.5%	13.8%	10.5%
	100.0%	100.0%	100.0%

[a]The figures represent the percentage of Southern Los Angeles County acreage devoted to each category of land use. They are based on the following sources: (a) Revised Land Use Acreages by Statistical Area, Los Angeles County Regional Planning Commission, April 1971; (b) Land Use Survey, Los Angeles County Regional Planning Commission, July 1940; and (c) 1990 Land Use, Los Angeles County Regional Planning Commission, September 1971. These figures were collected and summarized by Herbert Libow.

[b]"Used land" has minimal possibilities for future land development and may or may not have a current land use. Most land in this category is in use. "Vacant land" has possibilities for future development. Vacant land and used land are mutually exclusive categories.

2. The debate

Two policy makers – Adam and Bud – debate the "correctness" of the land use policy reflected in the above table by interpreting the data from their respective points of view.

Adam begins.

Adam: Well, our land-use policy as reflected by these land-use data looks pretty good to me. From the standpoint of regional and national goals it demonstrates much of the great accomplishment of man. We have identified great new resources and have been able to employ them for the betterment of our ever-growing population. Just look at the extent to which we have been able to expand our cities as reflected in the increased urban usage. This means that more people may have their own private homes and their own place to work. There is smog, crime and some congestion, to be sure, but we are making progress and our agricultural land is still adequate to satisfy food needs for some time to come.

Contained here in this single table is a message of great inspiration. We have discovered our earth, subdued it and put it to work for us. Truly earth is now resting comfortably under the dominion of man.

Bud: Oh, the treachery of it all, Adam. The data show that our policy has obviously been to consume and destroy our land and its resources through too much urbanization. Man can never really be superior to his environment, he's a *part* of it. What the data show is that we are in the process of breaking the basic harmony which man must have with his world. Man is such a recent arrival here on earth –

a few thousand years is nothing considering the billion of years that the earth has been around. Don't our predecessors, successors and partners have rights? Flora, Fauna and minerals have a rightful place in a harmonious world, too. We are offending these rights and may *not live* to regret it.

Sure, we derive a lot of pleasure and comfort today from our expansive sprawling urban areas, the products of our mines, our transportation system and so forth; but, the cost of depleting our environment is very high. We must learn to "give and take" with nature, take a little "bitter with the sweet", if man and this earth are both to survive.

Maybe now is the time for man to start looking inward a little more to find himself and his place in the universe. Earth is not an object to be conquered, but a *partner* with which to live. If man could only understand this he would learn to understand himself. And, after all, that's the real purpose in life.

I believe that 73.6% of our land in urban use is far too much. The projections are for it to get worse. Let's cut it down to 20% and let's cut down on the transportation usage as well. Return the rest to wilderness where it belongs.

Adam: Come on now, Bud. We have to *use* our environment to our own advantage. Nature changes the environment every day of our lives, doesn't she? Why shouldn't we change it? Make it work for us? Besides, I'll never understand your concept of returning land to wilderness. Do you mean that you want us to take large sections of land and just set them aside, never to be touched evermore?

Bud: Certainly.

Adam: How ludicrous. I agree with you that we must plan the ways in which we use our land very carefully. We have made mistakes in the past and chopped up wilderness areas much too fast and too rambunctiously. Certainly we aren't making the most efficient use of what we have. Some realignment and forethought are most assuredly needed. But to return to primitive times? Ridiculous!

In my view, the best land use principle is quite simple. Take a region of land. Survey it well and make a comprehensive plan for its use. Then make part of it accessible for man's use and part of it inaccessible. Don't touch the restricted portion until it's absolutely necessary. The 1990 projections seem consistent with this kind of planned land-use policy.

Bud: Would you preserve those restricted portions forever?

Adam: No, not forever. You can't take a large area of this country and keep it the way it was a thousand years ago. Nor can man afford to retreat from the most productive lands. We cannot continue to deny man's urgent needs by trying to create useless wilderness.

Bud: Wilderness is not *useless*. A wilderness area is worth saving or creating in its own right. Man has no business disturbing what lives there naturally. The pollution problem has shown that it doesn't take much to destroy ecological balance. In other areas our land use policy is literally driving some species prematurely into extinction. That's just not morally correct. They are partners in the universe. They have their rights, too. Only the forces of nature can sentence a species to extinction. It is not within man's moral purview to do so.

But, there is another reason for maintaining wilderness areas that perhaps you'll understand more quickly. Wilderness represents a bank for the genetic variability of the earth. It's our "gene pool", as it were. We are depleting this reserve at a very alarming rate. Now is the time to stop. What is wild, leave wild. Then let's take the land we've unwisely developed in the past and start a genetic investment program by rebuilding our wilderness "capital". Wilderness is a place where natural forces keep working essentially uninterrupted by man. Man should not interfere constantly in the process of natural selection and adaptation; for, the long-range survival of earth depends upon it. Earth always seeks an ecological balance. Everything is connected with everything else. Once man materially disturbs the balance the whole thing can go out of control. Puff! No more earth. Let's protect man from his ignorance now, if its not already too late. As Pogo says, "We have met the enemy and he is us!" The wilderness reserve is our best policy for protection against ourselves and our ignorance.

Man has taken enough for himself already. He needs to start paying back his debt. We should pretend that these wilderness areas don't exist. Don't ever disturb them. They are there for a different purpose – one that will insure the ecological variety that will ultimately save man. But that's not the only point. The important point is that we must learn to be humble with our land and its creatures and pay them respect. That's the proper thing to do; it's part of our obligation to maintain a harmonious universe.

Adam: While you're maintaining harmony, what are we going to do about the people? Population pressures seem to be irresistible. This population growth simply will not permit the kind of policy you're proposing. During the next 20 years we can expect the earth's population to increase from 3.5 billion people to a low of 5.5 billion people, perhaps a high of 7.5 billion people. The Southern California region alone (6 counties) is expected to increase from a present 10 million people to nearly 16 million in 1990 and 22 million in the year 2020. That's a lot of people to feed, clothe, house – to provide an acceptable standard of living. Where's it going to come from? Not from wilderness, that's for sure. It will have to come from the continued exploration and development of the land and eventually of the sea. We will have continually to discover and develop new resources in order to satisfy the demands of this ever-increasing population.

Bud: Just a minute, now. Those population figures are projections, not fact; I don't buy them. Indeed, the Washington Center for Metropolitan Studies suggests that there is a distinct possibility of achieving zero population growth within this century. We still have a *choice* in this matter. Population growth is not inevitable; in reality it's only an artifact of the extrapolations you make. But, if your crowd keeps talking that way, you'll come to believe it and then you'll go ahead and make it happen. Then we'll really be in trouble. The worst thing about all these population forecasts is that they tend to become self-fulfilling prophecies.

We have the knowledge and wherewithal today to control population growth and to level it off. It's naive merely to predict the population in the year 2020 and then plan how we are going to respond to it. The intelligent thing to do is to intervene now. We will have to intervene to re-establish the balance we have destroyed by excessive population growth. Within the limits of those already born, we can have any population we want in 2020. When we come to improve our understanding of the basic equilibrium of the universe, we will realize that this continued reproduction of man destroys the balance. Human population growth makes just too many of one resource-grubbing species.

Adam: Well I'm not as optimistic as you. History tells us that we just cannot put our "head in the sand" on this population growth issue. Any way that you look at it the species is going to reproduce. Furthermore, we know that human wants are insatiable. Elementary economics tells us that we are going to require more resources to satisfy the wants of this ever-increasing population.

This means we need a plan for earth-resource and land-use development. Resources are where you find them and their quantities finite. Perhaps the most criminal thing in the world is to waste resources or to let them stand idle when people need them. Our standard of living depends upon them. The trouble with your wilderness policy is that it doesn't take into account some basic facts about resources. Their original source is fixed by nature. So man must go to them, seek them out and develop them, no matter where they are.

Bud: Do you include the beaches, the Santa Monica Mountains and Lake Arrowhead? The Sequoia's, the Saguaro's? Grand Canyon, Yellowstone, the Tetons, Yosemite? King's Canyon? Lassen? Shasta? Lake Tahoe?

Adam: Yes, if necessary. If that's where the vital resources are. You have to go to the resources. They don't come to you. But they are the ultimate basis of our standard of living. That's where we get our foodstuffs, houses, furniture, television, telephones, electric lights, autos, and airplanes. Everything.

If we need lumber we should cut down those trees that provide the greatest yield for the cost of procuring them, but never more than we need. If that's the Sequoias, O.K. If Mt. Shasta has an economical deposit of an ore that is needed mine it. That ore may improve the standard of living for a hundred years to come.

Bud: Adam, you're all costs and no aesthetics. In the long run you'll have to pay for this excessively economic view of things. You know, there isn't much natural world left now. What there is, is *all* men will ever have, and all their children will ever have, too. Somewhere, sometime, we must learn to revere our sanctuaries of beauty. The most impressive thing about ERTS space photographs is that the natural magnifience of earth comes radiating through what is at best a cumbersome technology. But even now we can see the flaws, man's defiling of the splendor that once was the earth.

No Adam, your concept of value has you trapped. What's the worth of Yosemite? Fifty billion dollars? No! There's no price on it. The price of beauty has never been evaluated. Look at El Capitan! What would it cost to replace it? If we destroy it what is the "cost" to those who will never thrill to it, who will never know its beauty?

Adam: Bud, modern society has developed a very good method of arriving at such prices. It's called the market system. In it everything has a price – each commodity and each service – and is available for free exchange. Everyone receives money for what he sells and uses this money to buy what he wishes. If you're short of a good the price goes up, if you're long it goes down. This way an equilibrium between supply and demand is maintained.

Now producers, such as the mining companies, continue to compare the prevailing price of their inputs, such as ore, with the cost of its extraction and the market price of its final product by means of a profitability calculus. As long as the enterprise is profitable they are interested.

So, the time may come when Yosemite becomes profitable for ore mining. El Capitan will go into ore production if the mining companies outbid the other competitors for its use.

Bud: There's a limit to the utility of the market system when it comes to natural resources, a flaw which goes much deeper than just market imperfections or information lags. As you've defined it, only the producers bid, and only the *living ones* at that. We need to be sure that the aesthetic value, recreation value, peace-of-mind value, are cranked into these market decisions. In large measure these prices are beyond estimation.

But, it's not just the beauty, Adam, it's sustenance as well. The universe is infinite. We know nothing of its beginning or its end. There will (we hope) be many, many generations after ours. Where will *they* turn for the resources *they* need, when we have depleted them? And, there are the other creatures who inhabit and will inherit the earth. How does your market system take these other kinds of bids into account?

Adam: I am not going to penalize people today for the sake of future generations. Bud, my friend, "a bird in the hand, is worth two in the bush". Let me prove it to you. Suppose I offer you a choice between a dollar for certain today or a dollar a year, or say, a hundred years from now. Which would you take? The dollar today, wouldn't you? Why? Because the future is uncertain. Other things being equal, the longer we wait, the greater the possibility that conditions will change and the dollar will be worthless, or may never be paid back.

There is a second important reason why this is the right thing to do. The dollar received today can be invested to grow to more than a dollar a year from now, compounding as year builds upon year. The man who gets a dollar now and invests it wisely will have more money a year from now than the man who waits a year or more for his dollar.

Now, it's very much the same with resources. Their future value is less than their present value. An oil field, for example, is not worth very much a hundred years from now when we desperately need petroleum for energy today. For each year we fail to discover and develop new resources the world is losing some of this oil's value.

Bud: You're always thinking about just yourself, Adam, the all-powerful individual, the conquering hero. Never do you consider all of the others. You always place immediate needs ahead of those of the future generations and the other creatures. Remember that Noah took all species, not just man, on the Ark. He saw the need to preserve an ecological balance.

Consider, also, the needs of future generations. What's our obligation to them? Shouldn't the generations of the year 3000 "participate" in our resource decisions? 4000?

We should follow the advice of Theodore Roosevelt:

> The nation behaves well if it treats the natural resources as assets which it must turn over to the next generation increased and not impaired in value.

If we take him seriously then your idea of value is topsy-turvey. Resources are worth *more* in the future. We should at least be compounding forward rather than discounting backward. The future value of a resource is always higher than its present value. And, the "true value" of resources is very, very high. When we pump oil, mine ore or in any way spoil the beauty of nature, we change something *forever.* The recipient should at least pay a price equal to the cost of replacing the resource, of restoring nature to its original state; or he should pay the cost of depriving future generations of the resource.

Adam: Bud, those costs are unreasonable. They are astronomical.

Bud: Well, that's the true value. If you *really* need these resources, you'll be willing to borrow the money and amortize the debt over a few thousand years.

Adam, your version of the market system has made the cost of resources appear too cheap in the past. Consequently, we've become a society that is addicted to resource consumption. We had better "kick the habit" before it's too late.

Adam: Bud, you don't know what you are saying. If we place large amounts of land into wilderness and charge the fantastic prices you suggest, we'll be headed for an economic tailspin. Stagnation. Depression. Disaster.

The great accomplishment of the last 200 years has been our incredible ability to increase per capita real income despite an ever increasing population. This is, undoubtedly, Man's most singular achievement.

Bud: And, alas, it may be his very last!

Adam: Bud, economic development wasn't accomplished with the kind of mystical, religious, and fuzzy kind of thinking I've heard from you so far. No, it only was achieved through a rational, goods-centered concept of personal and social welfare. It was accomplished because we were dedicated to procedures for accumulating "factual data" and for measuring progress (such as the GNP). But, what a blessing the resulting economic growth has been for us. It has enabled the members of our society to look forward to an ever-rising standard of living. Economic growth, more-over is a great social solvent. When everyone can look forward to raising levels of well-being, the pressure for redistributing the existing pie is reduced. Bud, not to be in favor of economic growth is as heinous a crime in our society as is hostility to dogs and little children.

Bud: I note that you don't include the buffalo or the bald eagle. Was what we did to them in the name of economic growth a crime?

Adam: That's not the point, Bud. Economic growth is the way we satisfy people's wants and needs. Demand for goods and services is the basic force that keeps the system running. People want better meals, new clothes, more comfortable living, automobiles, T.V.s, washers, dryers, stoves. This consumption is made possible by the productive output of capital stock and the labor force, and through technical progress. As long as we increase capital and the labor force at a propitious and proportionate rate, and as long as we continue technical progress, we'll achieve full employment. Thus we will enjoy steady per capita income growth.

Bud: The theory of economic growth is doomed. We live on a finite planet with an infinite time horizon. Earth resource management must now take this into account. The rate at which our natural resources are being ravaged in our present frantic attempt to keep growing is frightening. As Thoreau said, "What is the good of a house if you don't have a tolerable planet to put it on?"

The trouble with technical progress is that it leads us down Ricardo's "road to oblivion". It increasingly forces us to use our more marginal resources. Take the case of copper, for instance. Early in this century if an ore wouldn't yield 2%–3% copper it was bypassed. Today 0.7% ore is mined as a matter of course. Soon it will be 0.35%. Then what? Will we dig huge open pit mines any place, anywhere an infinitesimal trace of copper is to be found?

This specter is only too clear today in the strip-mined hills of Kentucky. In the quest for coal to support our nation's growing need for energy we have literally denuded a good part of 11 counties of the state. Sure, strip-mining is less costly than deep mining and safer, in part because it employs fewer miners. But it has destroyed the terrain and led to erosion, land-slides, pollution, and, alas, now that the coal is nearly gone, economic depression.

And, there seems to be no stopping. The U.S. Geological Survey says that 3000 square miles in 26 states already have been "stripped", and that 71,000 more could profitably be gauged out in the future. But, it's not just the surface area that needs to be considered, it's the insides as well. Stripping has gone as deep as 700 feet in Germany. In the U.S. it could eventually go as deep as 2000 feet.

An alarming aspect of the L.A. county land-use data is that we are rapidly depleting our precious metal resources as shown by the decrease in tonnage. Meanwhile, our sand and gravel consumption goes up to construct our sprawling cities and expand our freeways. How much longer can we keep this up, Adam? When our mining resources and vacant land are gone, what then?

Adam: Of course, we must practice conservation to insure the maximum use of resources for the benefit of all people. But, stop exploration and development as you suggest? That's unthinkable.

There are some practical ways in which private enterprise can discharge its responsibility for conservation. Coals sells at about $7 to $15 a ton. Why not spend 5 to 15 cents a ton for general conservation and restoration? These funds could be used for developing and financing new mining techniques in which topsoil would be stripped off and saved, the minerals would be extracted, "clean" atomic explosives would be used to level the land, the surplus earth would be distributed and compressed, and the topsoil would be redistributed on the surface and treated with fertilizer. This program would be backed up by additional conservation policy to ensure that the land was reclaimable, that the vegetation would grow, that proper drainage patterns were maintained and that the atomic blasting was not hazardous to the residents of the area. In the future we will have to follow these conservation policies more strictly.

Bud: Conservation is humanity caring for the future, Adam, not technology and cost/effectiveness. This must become the age of man's service to man. Personal service can continue to grow with a static population and without destroying nature. We don't have to continue to think in terms of increasing production and consumption, supply and demand. We must come to realize *scarcity* is the nature of the world now, and forevermore. All resources are limited; nothing is supper-abundant. Man must cherish and live with his environment the way it is and *not* consume it excessively.

We must begin to take the "right view" of our world. All things are interconnected with everything else. Everything we do affects other things; everything goes somewhere. Pollution. DDT. They all go some place; affect something. In nature there are no "bottomless ponds". What is irrevocably destroyed or disturbed when we implement your plan, Adam? Are there not many unintended consequences from leveling the land, destroying habitats and valleys?

No, we must cease this craving for consumption. We must begin to live in a way that is least harmful to other living things, their habitat, their means of survival. In this way we will come to understand ourselves a little better, too, dispel some of our evil thoughts, arouse and maintain some good ones. The alert mind of the future must pay close attention to every state of the body, feeling and mind and attempt to place them in eternal harmony with the environment. This will require reflection and concentration. But the rewards are worth it. It is in this way we will find the infinity and pleasure we seek. Not an infinity of material resources, that doesn't exist; but, an infinity of freedom from excessive passion and desire.

Adam: Bud, it sounds like you are placing restrictions on man that even nature does not place on any carnivorous species. Man has a right to live, to eat, to shelter, to protect himself, and to grow.

Bud: Yes, but for the sake of harmony he should take only his share. These land use data indicate that he has gone too far. With the intelligence man has gained as a species there comes a moral obligation. He must strive to live in harmony with his environment. This means that our land-use policies must be changed.

Adam: Your plea is eloquent, Bud, but on the basis of the data and this debate I can only conclude that our past land-use policy has not been that bad. With intelligent planning for the future, and with effective world models, man will be able to make better use of earth resources and to continue his dominion over nature. The future of man is assured if he follows this approach.

Bud: No, Adam, from the debate one must conclude that our land-use policy has been disastrous. Man must stop thinking in terms of his dominion over the earth and begin to consider his moral obligation to all of creation. He must strive for harmony and ecological balance among all species and things. The hope of world models is that they may provide some insights as to how to accomplish this. Man's survival hangs in the balance. The future of man is *not* assured. It is in great jeopardy. We must adopt a new approach.

CAN WE MODEL THE WILD WORLD OR SHOULD WE FIRST TAME IT?

Claude FAUCHEUX, André LAURENT and Spyros MAKRIDAKIS
The European Institute of Business Administration (INSEAD)

1. Introduction

The last few decades have witnessed a widespread introduction of all kinds of models of social systems. During this period model building has grown from a state of infancy to full adulthood; and social scientists have been presented with a plethora of models. The advocates of modelling are full of euphoria. They insist that the human mind is not adapted to interpreting how social systems behave, a behaviour which is counter-intuitive anyway. They argue that mental representation of systems is fuzzy, incomplete and imprecisely stated and point out that the mathematics are the most powerful instrument man can employ in its quest to understand and predict social systems.[1] The critics, on the other hand, say that the potential of models is limited, while their present usefulness is nil. They seem to imply that models are an academic exercise aiming at elevating social science to a more respectable level than it is perceived at present.[2] Those in between tell us that models can be useful but we must understand their limitations. They point out their strong point but they also warn us of the dangers involved.[3]

In this brief paper we would like to reflect on the nature of models and explore their usefulness and applicability to social sciences. Our purpose will be to examine critically the adequacy and objectivity of modelling as a decision and policy making tool and suggest areas of improvement. Our aim will consequently shift to the future models and the changes which will be necessary to make them more responsive to the needs of generations to come.

[1] J.W. Forrester, Counterintuitive behaviour of social systems, Technological Forecasting and Social Change 3 (1971) 1–22.

[2] C.J. Crayson, Jr., Management science and business practice, Harvard Business Review (July–August 1973) 41–48, provides a comprehensive critic of management models, of the most developed and applied in the area of Social Science.

[3] J.C. March, Models bias in social action, Review of Educational Research 42, No. 4, 413–428.

2. The nature of models

Models are representations of real-life phenomena, situations or systems. They are abstractions of our environment since we cannot deal with its entire complexity. Models are not be any means a unique mathematical or O.R. activity. They are utilized widely by all living systems. Our perception of what surrounds us, for example, is only a representation of the environment we live in. In our minds we construct a model of environment because we are unable to deal with the complexity of the universe in its totality. This becomes clear by understanding how man's perception of what surrounds him has drastically changed over time, or by realizing the fundamental differences among various cultures, or even individuals within a single society, in the way they perceive the environment. There are even advocates who argue that environment becomes what we actually like it to be. There is no such thing as a predeterminated environment. Is is us who create one form or another to fit our own perceptive needs. As such, modelling is a highly subjective process, for two main reasons:

(a) there is no such thing as a "reality", directly accessible to our mind. On the contrary we apprehend the world by a process of interaction which develops and passes representations (codings) of the environments we live in.

(b) different actors will tend to develop models of the world differently because they have different experiences, needs, values and expectations.

Building a model requires a detailed analysis of the system we are concerned with. This is usually done through repeated observation so that the operational and behavioural characteristics of the system can be known. The outcome is a descriptive model of how we think the system may behave. It is an attempt to discover the critical factors affecting the system, how they relate with each other and to what extent they are influenced by outside inputs. However there is an infinite number of models one can develop to describe the same system. The choice of a single one depends upon the objectives of the builder, his abilities, the measurement tools he has at his disposal and the amount of time he is willing to devote in observation. However, even if all conditions are similar the outcome of modelling, attempted by separate individuals, can be quite different. This degree of subjectivity injected in the process of developing a descriptive model is over and above the perceptive one, already discussed.

A great deal of the current confusion as to the ability of models to describe, explain and finally predict the behaviour of a certain system came about from experience in the physical sciences where mathematical models were employed for such purposes. These models were more precise in describing the behaviour of the system than any of their verbal counterparts. They were easily manipulated so that variations in the system's behaviour could be explained. Finally, they were good predictors (to some degree of approximation) of future behaviour, the ultimate test of a good model. Engineers followed the physical scientists in building similar models; they even added a new type, purely predictive (e.g. time series models), without any descriptive or exploratory power. Consequently, several social scientists have thought that it is only logical to extend the introduction of symbolic models to their respective fields and therefore achieve the same results as with physical and engineering applications.

The argument usually advanced is that the mathematical language of symbolic models is definitely a more powerful tool of model design and manipulation than raw mental processes, a fact clearly proven in the physical and engineering fields. Furthermore, they say that it is a matter of time until advances in mathematics and computational abilities will enable us to deal with the bigger, more complex and less well behaved systems one encounters in social sciences. Thus, modelling has been attempted on a wide range of systems ranging from specific ones to "grand scale" encompassing the entire world system, such as the world dynamics approach of Forrester[4] and Meadows[5].

3. Problems with existing social models

Saying that we can transplant the modelling process of physical and engineering sciences to social ones is naive and can be dangerous. There is a major difference between the two: physical and engineering systems are non-evolving. Physical systems are in a steady-state of long-term equilibrium. Variations in their behaviour are perturbations around the equilibrium. Engineering systems are similar to physical except they are built to maintain a desired equilibrium. The designer analyses and synthesizes their internal structure so that outside disturbances will not alter the system's variables beyond the point of non-returning back to the equilibrium position. This is not true at all with social systems which are in a state of constant flux. They are in a kind of equilibrium but often a volatile one, changing its position, sometimes across extremes, a growth or evolving equilibrium one can see. Furthermore, the organization of social systems is quite different from that of physical or engineering. The latter are usually observed and studied, at the operating level. The former include higher level activities, such as adaptation, learning and self-organization in addition to the operating ones. The outcome is a completely different type of behaviour. This behaviour we think we can describe, we might possibly be able to explain at some aggregate level, but we are definitely unable to predict.

Let us take an example to illustrate the difference between, say, engineering and social systems. A computer is one of the most complex engineering systems. Given some electric voltage and a program, and assuming no mechanical breakdowns, its behaviour can be perfectly described, explained and predicted to the infinity of time. If we take, however, the determinant of all social systems, a human being, he

[4] J.W. Forrester, World Dynamics (Wright-Allen, Cambridge, Mass., 1971).
[5] D.H. Meadows et al., The Limits to Growth (Universe Books, New York, 1972).

will behave in a completely different way. We can give him a comfortable shelter and plenty of food and water, this will not keep him happy for long. He will soon learn to look for more mastery of his environment. Satisfying one level of needs will create another whose fulfillment can be even stronger than the previous.[6] He can forego the comfort and security of one job to go to another which provides more challenge and advancement. An "illogical" action, one can say, if he cannot understand the hierarchy of human needs. A computer will never deviate to such forms of "irrational" behaviour nor will a comet move away from its path because there are some complicated scales of needs and values to satisfy. The fact that it is only us and not natural, physical, or engineering systems that have a variety of never satisfied needs suggests that our behaviour is governed by higher levels of activities – activities which we just begin to grasp.[7] To say then that the mathematical models will explain and predict the behaviour of social systems because this has been done in the physical and engineering science misses the point because the systems studied by each one are basically different. Social systems have a highly complicated hierarchy of needs and values, they can learn, they have the ability to adapt to permanent changes in the environment, and they can even reorganize their structure to account for such changes. If we want to model them we will have to know how these higher level activities are performed. At present we can hardly discuss them on the descriptive level, and we are not even sure that such a description is correct. This is the main consideration which has been overlooked by many social scientists who try to equalize unequal things in their effort of borrowing the modelling concept from the physical and engineering fields.

The vigour and formalism of symbolic models can, no doubt, be helpful to social scientists once they change their approach to modelling to fit the needs of social systems. This may take some time, because of the required complexity but it should not result in a misdirection of efforts to something easier but not applicable. It is not formalism or mechanistic approaches we need but rather a new orientation which can deal with a novel problem.

Suppose for a moment that after a detailed analysis we can understand the exact behaviour of the stock market and furthermore that we can model it in such a way that future prediction can be made accurately. Such a model cannot be an adequate representation of reality once it becomes known. Obviously a large number of people would try to take advantages of its existence to their own interest. The result will automatically be a new pattern of behaviour in the stock market. As a matter of fact, some smart individuals will predict how others will react because they know and understand the original model. There will be smarter traders

[6] A.H. Maslow, Dynamics theory of human motivation, Psychological Review 50 (1943) 370–396.

[7] M.D. Mesarovic et al., Theory of Hierarchical Multilevel Systems (Academic Press, New York, 1970).

still who can attempt to guess the action of the second group. The stock exchanges can then change the rules of trading in order to eliminate the advantage one can gain because of the existence of the perfect model. The outcome will be a series of action-reaction moves going to, who knows, how many levels of interaction, whose effect will be to nullify an otherwhise beautiful theory. On our ex-post manner, one can go and trace each move in order to find what has happened and give a good explanation as to the reasons why it happened. But, in a predictive sense, we can do little because we cannot understand the intricate details of how millions of people will act and react when the conditions of the environment become different.

There is no necessity to examine fictitious illustrations; there are great number of actual examples showing how the physical or engineering approach to modelling has outrageously failed when social systems are involved. As a matter of fact to our knowledge there are no examples where models have successfully predicted the behaviour of social systems. About a hundred years ago Malthus developed a model according to which the starvation of manking was inevitable. Malthus was a good scientist for his days. He analyzed past trends carefully and he correctly discovered that growth in food was increasing linearly while that of population exponentially. This was an accurate explanation of what had happened. His mistake was the engineering or physics attitude that once we think that we understand the behaviour of a system at a given point in time, we can extrapolate it to predict its future. Well, we all know that the predictive ability of Malthus' model has been rather limited. On the contrary, production not only of food but other materials has been increasing in a much faster rate than that of the population. Other factors not foreseen at all by Malthus became critical. The rapid reduction in death rates, for example, has increased total population, not its fast reproduction. Production caused pollution, and material abundance created wealth distribution and social problems. It is these problems which create concern and not starvation per se. Today we can say that Malthus' model was naive. It took only two factors in consideration. He ignored technology, values, changing attitudes towards life, advances in medicine, urbanization and a plethora of other factors.

Unfortunately, today's models of social systems have not incorporated any of the criticisms one can advance about Malthus' doomsday fears. A main difference is that the majority of these models are mathematical and computerized. They are still extrapolation of past trends, present values and attitudes, with no attempt to incorporate changes, drastically new environmental conditions, learning, the development of new needs, etc. .... To become even more specific some of the better thought out social models, such as those of the systems dynamics approach, do not even include standard economic interactions, or basic human psychology principles. A well publicized conclusion of these models is that fossil oil energy will be depleted in the not so far future. It only took a limited war to create the energy crisis, change drastically the price structure, reduce energy consumption substantially, and induce a far reaching research programme for energy substitutes.

None of these has or could have been easily predicted. After all, there will be enough energy now for going much further into the future. This is a simple illustration which does not need to be expanded because the arguments for or against "grand scale" modelling have been publicized and are well known.[8] We want, however, to emphasize that the approach of "grand scale" modelling has not essentially been different from that of Malthus, that is an engineering approach applied to social phenomena. We then wonder if it will be possible to build another kind of model which could learn and adapt together with the social system it aims at representing.

4. Taming the wild world

In order to have a correct model we need powerful theories as well as accurate and consistent data. The former is completely lacking in social sciences where theories often verge on speculation, while the latter is non-existent.[9] But even assuming that correct model could exist, this would not have solved the problem of their application for decision of policy making. March[10] expresses this clearly when he says: "different models suggest different actions, and the attractiveness of the social and moral consequences of those actions do not depend entirely on the degree to which the models are correct. Nor is this problem solved in any significant way by producing a more correct model." Is then the modelling approach the right answer to explaining or predicting the behaviour of social systems? We believe that it is naive to model the world as if it were a mechanical object. Instead, we need to tame the wild tendencies in us. The problem, so far, has been our incapacity to anticipate the long term consequences of our actions which must be understood and assessed in their totality before we move to model the world. From all beings we are the only ones with the dual capacity of both actor and designer: our needs determine our present actions, our values influence our expec-

[8] D. Kild, (Computerized prophecy: The limits to growth, Contemporary Psychology, 18, No. 1 (January 1973) 1–3) summarizes the criticisms as (a) the simplistic attempt to treat the entire earth as a single system (b) not taking into account changing human behaviour (c) the naive assumptions of static limits of food and resources (d) the especially weak assumption of catastrophic rise in pollution (e) the complacency of those who have "made it" about a no-growth world, and (f) the failure to present the data on which the apocalyptic predictions are made. In terms of advantages "grand" scale models have made crystal clear the dangers of growth, and have initiated considerable debate among scientists policy makers and common people.

[9] MIT, Man's Impact on the Global Environment (MIT Press, Cambridge, Mass., 1970), put the lack of data as follows: "We found that critically needed data were fragmentary, contradictory and in some cases completely unavailable. This was true for all types of data – scientific, technical, economic, industrial and social" (p. 6).

[10] March, op. cit., p. 414.

tations, while the needs and values of others may represent constraints for the satisfaction of our needs.

Unless the dynamics of needs, values and sharing environmental resources can be understood, we could not assess our role either as actors or designers. Furthermore we cannot predict such role and therefore say how social systems will behave. A "black box" approach cannot work when the actor is at the same time the designer. If the need for material possessions and profit, for example, saturates or decreases (a phenomenon already appearing in certain groups of people) its repercussions on the future can be completely different than what we envisioned it now. For once the burden on natural resources will diminish; furthermore if this is coupled with a better product design aiming at minimizing obsolescence (assuming a lesser desire for profit) the effect will be more drastic. Finally, if technology is taken into account which could discover new and better methods of production we can even assume an abundance of material resources to the point that everyone in need can easily obtain them, that is a switch to a system where supply will exceed demand and price reduced to almost zero. Of course, other assumptions about the values to prevail at some future time will result in different scenarios as to what the future would look like. What we are actually saying is that man is not a passive participant of history; he rather creates it through his own wants and values and he is increasingly inclined to actively participate in the shaping of the world.[11] It may therefore be better not to try to extrapolate in order to predict the future but rather intrapolate from some future point to the present. That is, models could be more accurate representations of social systems if they were teleological rather than exploratory. A fact that will require a new discipline of social forecasters, mainly experts in human psychology, sociology, history and political science who will get together to determine how values will evolve and what kind will prevail in the future. Next step afterwards would be for the mathematicians and O.R. experts whose job will be to formalize values into their models of social systems. In the absence of value within the models, prediction cannot make sense.

Another factor of importance is that of constraints. Constraints do exist in the natural, social and cultural environments but are perceived differently by various people. What Westerners consider minimum standards for housing and nourishment are luxuries for the majority of mankind. At present it is the Western, industrialized societies which mainly determine what to be produced, set prices, and in many ways, decide how wealth should be distributed on a world wide basis. It seems that

[11] N. Johnson and E. Ward, Citizen information systems: Using technology to extend the dialogue between citizens and their governments, Management Science 19, No. 4, Part 2, P21–P38.

this system of wealth distribution could not last for long.[12] We have already seen developing nations to be asking for a fair share of materials goods. The energy crisis is only the beginning, other nations, rich in raw materials will follow, demanding higher prices which will eventually result in a more fair wealth distribution. The effects could be sweeping, changing radically wealth patterns. What will exactly happen is another area needing examination but its effects cannot be ignored. However, it seems that a more equalitarian world will eventually emerge. The question of how this will happen deserves a closer look because it is going to be a major source of conflict which together with that arising out of East–West competition will greatly influence the future shape of our world.

Our ability to resolve conflict among ourselves is an art we know little about. Going to the extreme and assuming complete mastery of conflict through effective social debate among, say, nations at some future time would release an unpreceding amount of resources which could be devoted to satisfying social needs.[13] On the contrary an arms escallation battle can only alleviate existing problems. How conflict resolution will evolve in the future is another area requiring close scrutiny by teams of social scientists who will have to determine the intricate relationship among needs, values, constraints, perception, and conflict resolution. It is this type of explanation which will be extremely useful while constructing better models of social systems. Unfortunately the interrelationship involved is not obvious, not linear and much must be done to discover it, and it is only in the process of designing such interrelationships that we will begin to understand them. In order to achieve this, we will have to realize that designing the world of tomorrow will be an on going task involving all participants of change as the designers. This means that social debate will have to develop in many places at many levels. We will have to learn better how to solve conflict through debate rather than through violence. We will have to learn from other cultures and civilizations which might have more to offer than we have been willing to recognize and accept. Civilization is the process through which violence is transmuted into debate. Before we can think of resolving the violence we do to nature we will have to reduce the violence among us, which ultimately means taming aggression in ourselves. This will involve making very explicit what we *want* to become, what we *could* become and

[12] J. Reston (Impeachment at the United Nations, Herald Tribune (April 19, 1974)) talks about the inequalities in wealth distribution between the rich, developed and mainly western countries and the poor, developing and mainly non-western nations. He characteristically says the "Industrial West . . . are cheating the poor countries by buying their products cheap and selling yours (West's) dear. This unequal exchange between the cost of labour in the rich nations and the poor nations is the heart of the problems".

[13] K.E. Boulding (The Economy of Love and Fear (Wadsworth, Belmont, Calif., 1973)) asserts that "The world war industry as measured by military expenditures, is of the order of $ 200,000,000,000". He further says "It is often unrecognized paradox that the cost of maintaining a war industry is usually much greater than any actual damage to an enemy it can cause" (p. 80).

what we *should* become. This will require a more effective organization of values transactions among people and across cultures.

In concluding we would like to point out that the Earth as a natural and social system, has succeeded in surviving for several billion years. One can of course, say that there are always first time occurences but these are highly unlikely: catastrophes do not come in complete destruction patterns. Before a mass starvation there must be a limited one which will reduce world population and increase the per capita food. Before a catastrophic pollution there will be a heavy smog in big cities, if this becomes too unpleasant enough people will move out to the country a fact which by itself will reduce the magnitude of pollution. Also there will be laws and controls to alleviate the problem. You cannot, therefore, fly to extremes, you have to reach there slowly and usually there will be time to learn and take corrective actions if necessary. It is the actor/designer role which makes the difference. Unfortunately, existing models do not take such simple and logical considerations into account. Doomsday predictions make better headlines than good news.[14] Furthermore, they can be commercially more profitable when public awareness about certain topics, e.g. ecology, is high. But one has to distinguish between good science, necessary art and successful journalism. The latter is tempting, while the former is difficult and essential.

[14] It is like the movie "Exorcist" whose phenomenal success is based on territorizing its audience.

THE LOGISTICS OF SIZE

Martin K. STARR
Graduate School of Business,
Columbia University

Each organization has a best size. The problem is to find the variables that determine best size and their relationships. It may be that the variables and the functions are not dissimilar for profit and public systems. Thus, a hospital and a pretzel manufacturer may have more things in common with respect to their optimal size than one would initially suspect.

In this paper, three major variables are explored with respect to their effect on size. First, is productivity (production efficiency is an internal cost). Second, is the cost of marketing against competitors (an external systems factor). Third, is management costs (an internal cost related to information flows and organizational structure). Comparing public systems and private ones, it is possible to substitute for the second cost listed above, a cost of variable demand (an external cost of uncertainty). It has much the same functional form with respect to size, S, as the costs of marketing against competitors has. The best size for a hospital, library or electric power company will be determined in part by the swings in demand from peak to peak.

Before developing the model for organizational size, let us dwell a bit on what we know about the size of organisms because the analogy with organizations seems fitting, although subtle.

1. For organisms

There is an optimal size for any organism which is in accord with its structure and its environment. Evolutionary process continually shapes the organism in response to fluctuating conditions searching for the optimal state. But as in the case of the dinosaur, the rate of environmental change can outspeed the natural process.

D'Arcy Thompson in volume I of his great text *On Growth and Form* (Cambridge University Press, second edition, 1942)[1] discusses many kinds of size relations. In his second chapter, "On magnitude", he states (p. 24) "Everywhere

[1] First edition, 1917.

Nature works true to scale, and everything has its proper size accordingly. Men and trees, birds and fishes, stars and star-systems, have their appropriate dimensions, and their more or less narrow range of absolute magnitudes."

Absolute magnitudes become relative or proportional magnitudes when the organism is modeled in terms of the environment (e.g. effects of competition or fluctuations in supply and demand of necessities). Only by considering relative magnitudes, can we develop models that are appropriate for determining optimal sizes. Thus, in biological systems, where gravity is a major environmental factor, the weight of the organism (W) is generally proportional to the cube of its length (L), i.e.

$$W = kL^3$$

where k is a constant characteristic of the species. Gravity, as an environmental force operates on the cube root of weight as a constraint to length. Doubling length requires an eightfold increase in weight.

The thickness of limbs and stalks is related to their cross-sectional area. Quoting from D'Arcy Thompson (p. 26) "... since the weight of a fruit increases as the cube of its linear dimensions, while the strength of the stalk increases as the square, it follows that the stalk must ... grow out of apparent due proportion to the fruit ... and yet again, that in quadrupeds a large head must be supported on a neck which is either excessively thick and strong like a bull's, or very short like an elephant's."

Climate is another force that acts on biological size, e.g. environmental temperature. Noting that the volume (V) to area (A) ratio of a sphere is $V/A = \frac{4}{3}\pi R^3 / 4\pi R^2 = R/3$ and that for small animals, $R < 1$, Thompson states (p. 26) "... the larger ratio of surface to mass in a small animal would lead to excessive transpiration, were the skin as "porous" as our own ... we may thus account for the hardened or thickened skins of insects and many other small terrestial animals."

Similar examples of size relations can be drawn for inert, physical structures such as buildings and bridges. Strength of materials under different forces have been studied at length to allow design optimization of form and size. When a new material is developed or a new structural concept appears, size constraints change. Factors that constrain organizational size would also be subject to change, as both internal and external elements alter. For example, the computer lowered (internal) information processing costs and thereby increased the optimal organization size. This will be apparent from the model that follows.

2. For organizations

It is easy to accept the notion that organisms evolve to an optimum size. It is reasonable to expect that buildings and bridges are designed to an optimum structural size. Is it less apparent that there is an optimal size for any organization?

If there is an optimal size for any organization, then that size specifies the limits to growth. Conversely, it indicates where further growth should occur. Put in this way, the question of growth can be answered by rational analyses based upon the logistics of size.

An appropriate model must be developed to identify the forces that affect organizational size. These forces will be analogous to gravity and temperature for the organism; and to tension, compression and torque in physical systems. Also, the appropriate characteristics of the organizational structure (analogous to area, volume, etc.) must be found. Our models should capture relationships that are appropriate to organizational size – much as for organisms such features as volume to area ratios, surface textures and branch thicknesses are shaped in response to the forces of gravity and temperature. We note that neither organisms nor organizations are simple systems whose size is completely determined by one or two straightforward relationships; but as with organisms, certain forces may play a major role in determining size.

In this paper, an attempt is made to isolate the costs associated with three critical factors (productivity, competition and/or uncertainty, and administration) but major research efforts are warranted to get at all relevant variables. Not much is known about the forces and functions and tradeoffs that determine optimal-sized organizations and optimal-sized divisions of such organizations. If they were known, then rational regulation of size . . . to maximize the benefits to society . . . could replace clumsy anti-trust procedures. A rational policy of regulation would allow society to know what its laws affect and in what way. A great deal of interest would develop in assessing the optimal size of multinational and international organizations. The nature of conglomerates would be better understood and the regulation of public utilities might be better tuned. It is appropriate that management science develop its resources to make this contribution because it alone has the systems capabilities to do so.

3. Determinants of organizational size

3.1. Productivity

In a specific market with n competitors, generally, the larger the competitor the greater its production advantage. The amount of advantage will differ according to the process. The result reported by The Boston Consulting Group[2] is approximately a 25% decrease in unit costs for each doubling of the production output, thus:

$$P_S = P_1 \left(\tfrac{3}{4}\right)^{\log_2 S}$$

[2] Publications available from the Boston Consulting Group, Inc., 1 Boston Place, Boston, Massachusetts 02106, USA.

where P_s is the unit production cost at size S and P_1 is that cost for the minimum organization size.

There are numerous reasons why this kind of relationship might hold (not concerning ourselves with whether $\frac{3}{4}$ is the appropriate value for the rate of cost decrease). Flow shop design (continuous production) can be used in place of job shop design (batch production) when S is large enough. Additional experience with production gradually fashions an ever-better flow shop. Furthermore, better discounts can be obtained with larger purchase volumes, buy decisions shift to make, and faster equipment is warranted. Modularity and group technology may provide additional cost reductions.[3]

3.2. Competition

Now let's consider unit marketing costs. They are likely to increase at a diminishing rate with increasing values of S . . . for the following reason. A market that is in equilibrium will have an equal number of customers switching between any given pair of brands. Fig. 1 illustrates this situation. (Note, for a more general interpretation of the situation, this can be restated as follows: A system of alternatives that is in equilibrium will find an equal number of persons switching between any given pair of choices.)

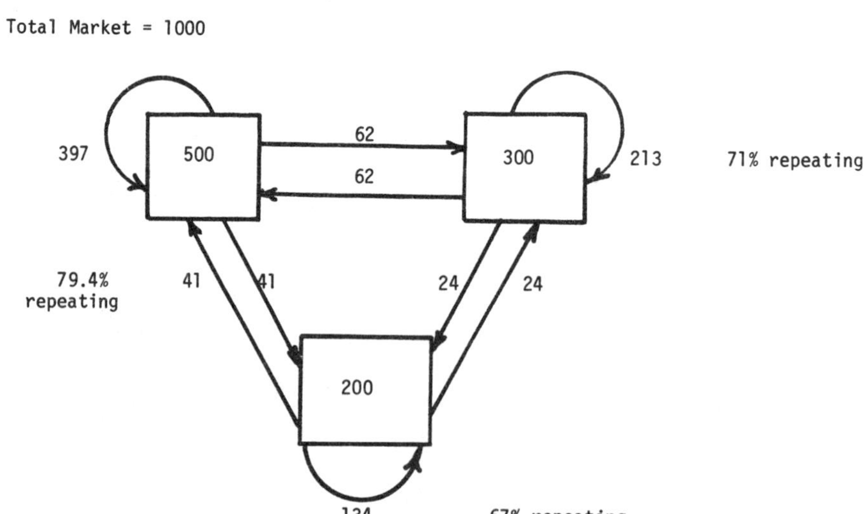

Fig. 1.

[3] Martin K. Starr, Modular production – a new concept, Harvard Business Review 43, No. 6 (November-December 1965) 131–140.

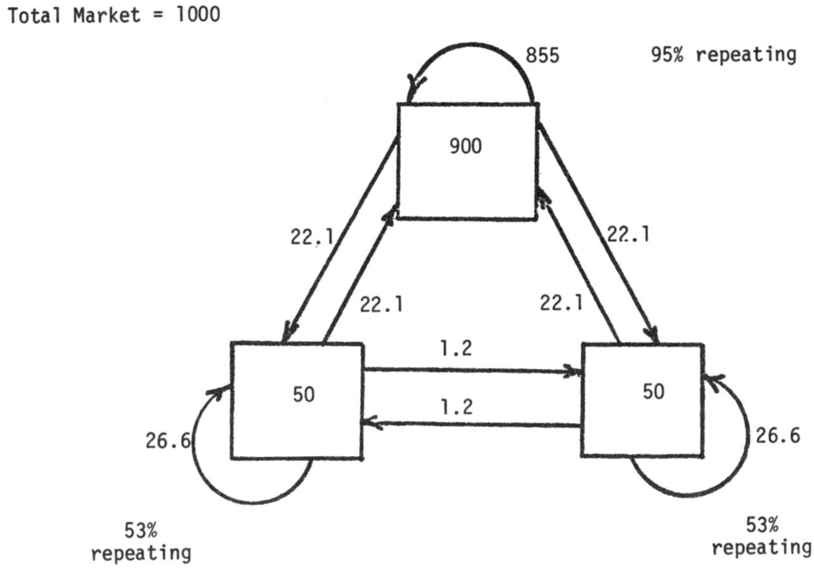

Fig. 2.

There is considerable empirical evidence to support the notion that most short-interval decision sequences (or short purchase cycle) markets are in equilibrium. But even when there is some deviation from equilibrium, big share brands must obtain greater loyalty, i.e. maintain a higher percent repeating. Note that in the prior example, the 50% share brand is required to generate 79.4% repeat if it is not to shrink in size, whereas, the 20% share brand needs only 67% repeating.

An even more exaggerated case is that of the three brand configuration: 90/05/05 (fig. 2).[4] In this instance it is noteworthy that even quite small organizations (5%) must maintain more than 50% repeat. The giant organization with a 95% repeat requirement is likely to have to spend more marketing dollars per unit repeating than the 5% brands. Therefore, marketing unit cost may be thought to change with size (S) as shown in fig. 3. If this is reasonable, then unit marketing expenses rising with size act as a constraint on the growth of ever-larger (and more efficient) production systems.

Another way of looking at this matter is in terms of holding costs and switching costs. Say that a salesforce is employed to get new customers and to hold old customers. What mix of these two activities should be used? A small

[4] The figures for switching and repeating used are based on the Hendry model for detailed balancing. (See David H. Butler and B.F. Butler, HendroDynamics: Fundamental Laws of Consumer Dynamics, Chapter I, 1970 and Chapter II, 1971, The Hendry Corporation.) The same conclusions would apply however, even if somewhat greater or lower switching levels were employed. The general size of these figures has been substantiated by empirical studies.

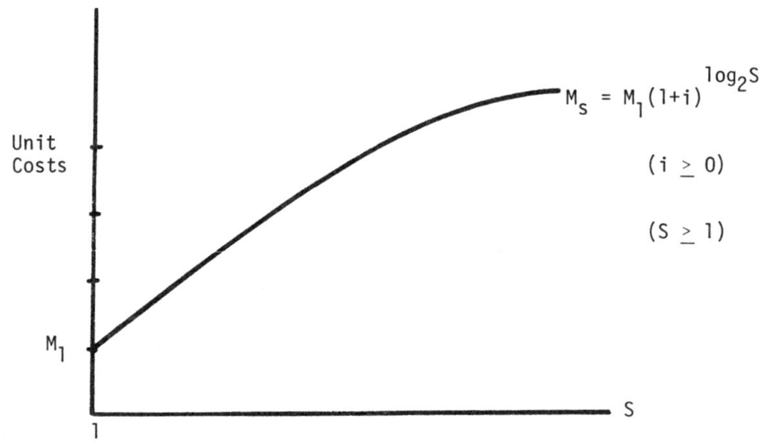

Fig. 3. M_S = unit marketing cost at size S; M_1 = unit marketing cost for the minimum size organisation; i = holding cost parameter.

organization should reach a different answer than a large one. If the costs of obtaining switching are equal to those of holding, then the per unit marketing costs or organizations of any size are equal. However, if the holding costs are larger than those of switching, the differential becoming greater as the per cent of the total to be held increases, then the per unit competitive costs will rise in the manner shown above.

The model we have developed so far is applicable to an organization which, first, produces some goods or services where increased production output improves productivity, and which, second, has competition for its goods or services reflected by a sequence of decisions to repeat or switch alternatives and which, third, requires greater expense to maintain loyalty as its share increases.

4. Uncertainty

In the case where competition is not relevant, e.g. a public service facility, there is an increasing cost with size (S) which is comparable to the competitive cost just discussed. Assume that two organizations of equal size s provide non-competitive service to an equal number of users. Let the variance of demand for each organization's output be σ_s^2. Then if these two organizations were merged to form a single system of size S, the total variance of demand for output would be

$$\sigma_s^2 + \sigma_s^2 = 2\sigma_s^2$$

whereby

$$\sigma_S^2 = 2\sigma_s^2 \quad \text{and} \quad \sigma_S = \sqrt{2}\sigma_s.$$

If costs of providing service are relatively proportional to the demand uncertainty, then, in general

$$\sigma_S = \sqrt{S}\sigma_s$$

and the costs increase with size in much the same fashion as the competitive costs previously discussed. Given that the cost of uncertainty rises with size, this effect acts as a constraint on the growth of increasingly efficient production systems.

In the material that follows we will examine the total unit cost function for productivity and competitive costs. The same type of analysis could as well have been done substituting the costs of uncertainty for those of competitive marketing.

5. Minimum total unit cost

When the production function is added to the unit cost marketing function, there is clear indication that a *minimum total unit cost* can be found which would be associated with an optimum size for the organization (fig. 4). This optimum is predicated on zero excess cost being passed along to the consumer. The producer has maximum margin to work with, but not necessarily maximum profit. If additional output is required, then a minimum total unit cost solution would be found subject to the demand constraints. While it is possible to redefine the optimum organization size in terms of maximum profitability, it is not in keeping with our intentions to do so.

The mathematical expression for minimum total unit cost is given below, with some observations concerning the variables.

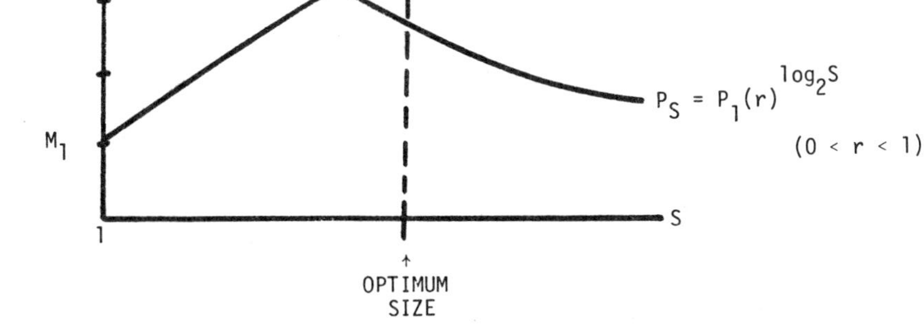

Fig. 4.

The minimum value of total unit cost (*TUC*):

$$TUC = M_S + P_S = M_1(1 + i)^{\log_2 S} + P_1(r)^{\log_2 S} \qquad (i \geqslant 0, S \geqslant 1, 0 < r < 1)$$

is associated with S, as follows:

$$S_{\text{opt}} = \exp \left\{ \left(\ln \frac{-P_1 \ln r}{M_1 \ln(1 + i)} \right) \Big/ \log_2 \left(\frac{1 + i}{r} \right) \right\}$$

Ceteris paribus, the optimal size, S increases when

P_1 increases	M_1 decreases
r increases	i decreases

Thus, the *optimal organization size is large* when:

- base production unit costs are substantially greater than base total marketing unit costs ($P_1 \gg M_1$) e.g. raw materials converters;
- productivity improvement with increasing volume is quite small – e.g. capital intensive products as compared to labor intensive products (a technological discovery can convert the latter to the former thereby opening the way for a large, dominant organization to emerge);
- unit marketing costs decrease – they do not have to decrease a great deal to sustain a significantly larger share of the total market – e.g. basic commodities.

These observations help to classify what kinds of industries would tend to be larger or smaller, but they do not shed light on the sensitivities of size to changes in the values of the variables. This subject will be discussed briefly in the next section. Before, turning to sensitivity analysis, however, let us consider the availability of empirical values for the modeling parameters.

According to The Boston Consulting Group, r is often to be found in the neighborhood of 0.75. Knowledge about holding and switching costs is rudimentary and must be developed. Even the shape of the curve for marketing unit costs is hypothetical and must be verified by experimentation. Nevertheless, it does seem possible to obtain the kind of data that this model requires.

The costs of uncertainty could be developed by type of organization and size. The lines for designing adequate research projects are available.

6. Sensitivity analysis

Some tests of the properties of the above model were made using $r = 0.75$ and $i = 0.10$ as a base. Then small changes were made in these values to observe the effects on the optimal value of S. As expected, when the unit cost of holding was increased, the optimal size of the organization decreased. For our simulations, the decision concerning size appeared to be highly sensitive to the holding cost parameter, i.

When the unit cost of producing decreased, i.e. more than a 25% decrease occurred in unit production costs with each doubling of output volume, the optimal value of S decreased – but not by a great deal. The response was relatively insensitive. This point can be explained by the fact that the production unit cost curve becomes quite flat at large output volumes.

We note that the majority of organizations in our essentially job-shop oriented world operate with small output volumes that put them in the high unit production cost portion of the production function. This is especially true of public service organizations. Although their market costs (or costs of uncertainty) are lower as a result, their place on the total unit cost curve identifies them as marginal producers. This translates into higher prices and lower product qualities and service performance for the ultimate consumer.

7. Management costs

Managing the organization entails significant costs that vary according to the size of the system. Many complex factors interact, in this regard, reflecting the amount and types of information flows, organizational structure and leadership. But these factors are so highly interrelated that a reasonable hypothesis is to reduce the apparent complexity of the system, according to its degree of centralization, by allowing management costs to be a simple function of organizational size.

In a highly centralized bureaucratic operation, managerial costs increase exponentially as the organization grows. (This is based on the fact that S nodes are connected by $S(S-1)$ arcs.) Thus, administrative costs for large-size organizations would be *proportional* to S^2. As the degree of centralization decreases, subsystems of sizes S_1 and S_2 appear, where $S_1 + S_2 = S$. Consequently, with decentralization, management costs decrease since:

$$S_1^2 + S_2^2 \leqslant S^2 .$$

Yet, no matter what the degree of decentralization, following this model, management costs rise exponentially, faster or slower. (The exception is the anarchistic case where $S_i = 1$ for all i and since no one speaks to anyone else,

$$\sum_i S_i(S_i - 1) = 0.)$$

We also consider the possibility that the costs of managing increase at a decreasing rate. The model, in this case, would be similar to the one derived for competitive costs, and also for the costs of uncertainty. Let I_s = the cost of managing a system of size S where I_1 = the basic costs for managing an organization of minimum size and where j is the cost parameter of the particular organization embedded in a specific institutional setting or industry. We may then write for this cost:

$$I_s = I_1 (1 + j)^{\log_2 S}.$$

We are familiar with the effect of this kind of cost component. The size of the optimal organization will be reduced as j increases, because larger j means it costs more to handle transactions. The increased costs reflected by larger j do not have to be restricted to actual payments for work. Instead, they should be thought of as encompassing employee satisfaction, benefits of job enlargement and worker morale in general. Optimal size also will decrease when I_1 increases which is reasonable because large I_1 indicates a high initial fixed cost for managing the system.

The effect of the exponential formulation of the costs of managing (i.e. proportional to $\Sigma_i S_i^2$) is to require an even smaller organizational size than the model using the log exponent ($\log_2 S$). In either case, communication requirements act as a constraint on organizational growth if the objective is to minimize total unit costs of output. The constraint on size is greater when more bureaucracy prevails.

8. Distribution costs

There are some interesting variants of organizational form that can, under specific circumstances, alter the optimal organizational size. For example, if a large producer (having low production unit costs) can achieve broad distribution at a low per unit distribution cost, then the benefit of being No. 2 in each market (the local, regional producer is No. 1) is lower holding costs. See the table below.

Table of shares.

Brands	Regions		
	1	2	3
A	40[a]	10	–
B	20	–	60[a]
C	10	50[a]	–
X	30	40	40
	100	100	100

[a]Regional producer.

If sales are 100,000 in each region, then X is the largest producer with volume of 110,000 but it has lower marketing costs in each region. This marketing advantage results in an even larger optimal size being indicated for this already efficient producer. The same kind of reasoning could be applied to multinational organizations in order to discern what are their strengths and weaknesses.

It is also possible to recast this model in terms that apply to non-competitive situations or to where demand uncertainty costs prevail. The central warehouse model[5] captures the effects of different distribution configurations on costs.

9. Monopoly

The model we have been developing calls attention to the fact that a monopoly can operate much closer to the optimal production size because it has no holding/switching costs to contend with.[6] But even with monopoly conditions, size is still constrained by uncertainty and by bureaucratic costs. As bureaucratic costs can be controlled, greater production efficiency can be obtained. This objective should be pursued eagerly by students of organization design and by management practitioners as well. The computer can be used to break out of bureaucratic patterns if it is not employed to copy what has always been done before . . . but faster.

Especially, with respect to international trade the monopoly form has great appeal. The optimal producer will have a very high volume production system. No aggregation of small producers, even if operating through a trading company or with governmental help, will be able to match the performance of the optimal producer. It should also be noted that the costs of uncertainty can become proportionally less for the large multinational firm if it utilizes a central warehouse type of configuration.

The term monopoly connotes bad things. It should be dropped. Laws governing size should be redrafted. Growth in organization size is too easily associated with good things. Management studies should teach it otherwise. Optimal size is our objective.

[5] See Martin K. Starr and David W. Miller, Inventory Control – Theory and Practice (Prentice-Hall, Englewood Cliffs, N.J., 1962).

[6] The weaknesses of monopolies (e.g. overcharging, inhibiting technology, etc.) do not appear in this model but they can be overcome with rational regulation, which will increase management costs.

BACKWARD INTEGRATION TESTS OF DYNAMIC MODELS*

Richard D. WRIGHT
School of Industrial and Systems Engineering,
Georgia Institute of Technology

Backwards integration tests simulate models backwards in time in order to compare retrospective projection with historical observation. The general relevance of such tests for assessing the validity of dynamic, social system models and the particular applicability of backcasting to the Forrester World Models are topics of continuing controversy. The controversy is resolved by establishing four points: (1) Backward integration is an uncertain method of exercising dynamic models; spurious results may easily be generated. (2) Not all backcasts are spurious; the explosive behavior observed by backcasting Forrester's World Model behind 1900 is computationally correct. (3) The numeric instability of backward integration does not preclude testing model behavior for pre-initial periods; retrospective comparisons can be accomplished directly by forward integration. (4) The interpretation of non-spurious backcast failures depends on a model's range of applicability; the pre-1900 misspecification of *World Dynamics* is not relevant to that model's operation after 1900.

1. Introduction

Open debate is one forum for confronting and improving complex public policy models. An example of this process appears in *Models of Doom* [5], a detailed critique of the Forrester World Models[1] and a rebuttal by the authors of *The Limits to Growth* [10]. Unfortunately, a technical argument about the feasibility and utility of "backcasting" the World Models has obscured much of the thoughtful and accurate criticism of those models' scope and content contained in *Models of Doom*. Beyond the context of the World Models, an analysis of backcasting is of interest for two reasons. First, backcasting has been proposed as a generally useful validation procedure. Second, if in fact dynamic models of societal systems are prone to spurious explosions in backward evolution, then it will not be possible to apply some of the commonly used optimal control algorithms to those models.

Backcasting tests integrate dynamic models backwards in time and compare retrospective simulations with historical observation. Controversy about backward integration is due to two explosive backcasts of the original world model, For-

* The author thanks Professor Philippe Wilmes and Mr. Jerry Richards for their helpful comments.

[1] The "Forrester World Models" refer to Forrester's "World2", documented in *World Dynamics* [6], and its successor "World3", described in *The Limits to Growth* [10].

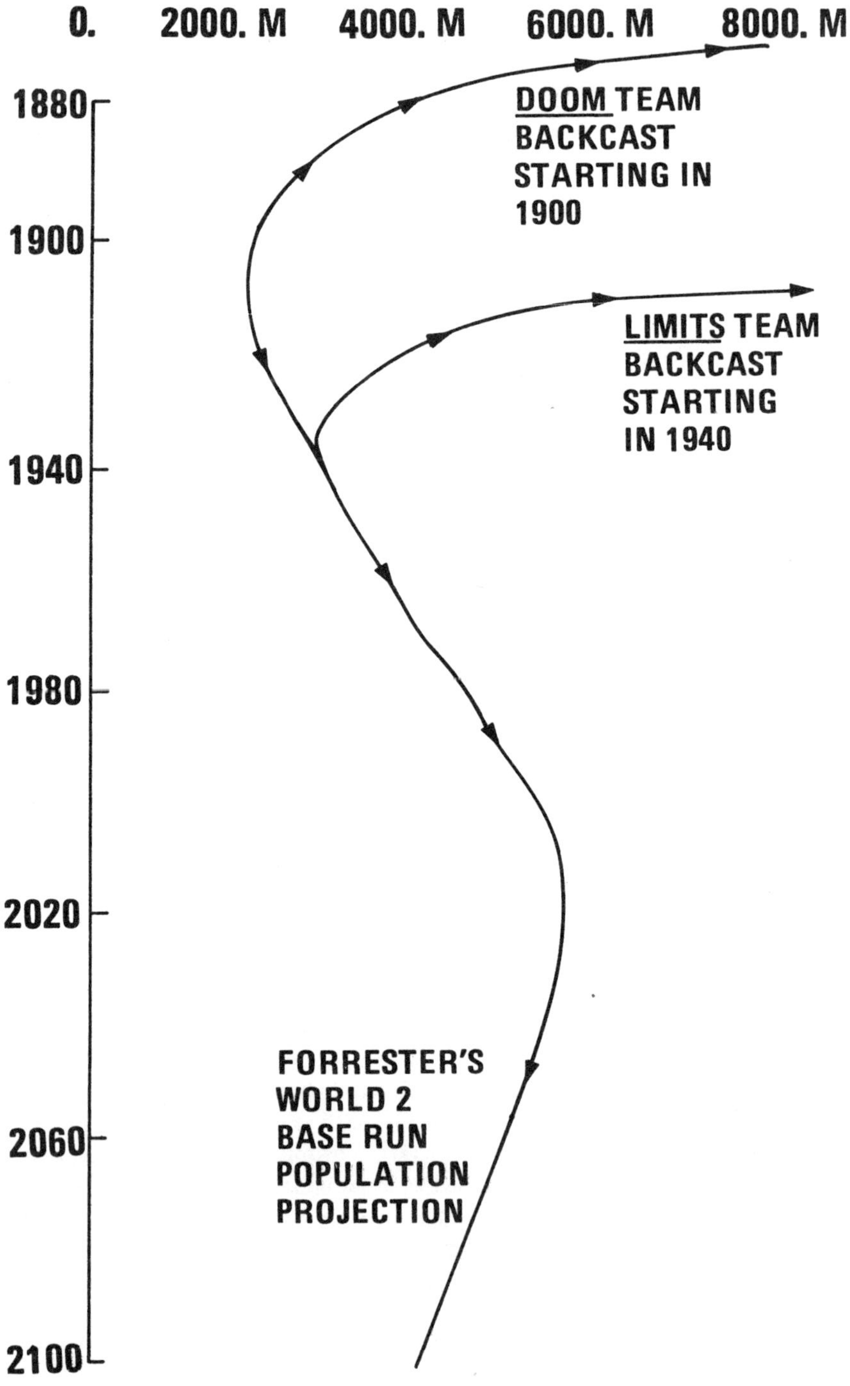

Fig. 1. Two backcasts of the *World Dynamics* model.

rester's "World2" [6, p. 132]. These two backcasts together with Forrester's forward trajectory for population are shown in fig. 1. Playing the World2 model backwards from 1900, the *Doom* team discovered grotesque results:

> "As it stands Forrester's World2 will not run backwards from 1900 for more than a little over 20 years . . . the 20th century lies in the aftermath of a catastrophic population collapse . . ." [5, p, 113].

The discrepancy between the world population value found by backward integration and estimated 1880 world population (and a similar finding for the *Limits to Growth* model [5, p. 116] caused the *Doom* team to question the utility of both World models:

> ". . . this must throw a certain amount of doubt on the validity of the results and any conclusions drawn from them about the future" [5, 117].

On the other side, the *Limits* team point out that the World2 model explodes when integrated backwards from its own 1940 values (shown above in fig. 1). The *Limits* team argues since "World2 does not always backcast its own behavior . . . running a systems dynamics model backward tells us nothing about the model's utility in understanding the world;" [5, p. 221]. The Limits' rebuttal then asserts that the Sussex result is "completely spurious." [5, p. 222].

The *Doom* team, in response, argues that backward integration is both feasible and useful:

> ". . . It is only reasonable to expect that a model should "backcast" or extrapolate into the past . . . To argue that one should not attempt to backcast using a System Dynamics model, while at the same time admitting that the same mechanisms operated in the past, is to say that such models are beyond the reach of normal scientific criticism" [4, 64].

The utility of backward integration as a general validation test and the particular results of the *World Dynamics* case are disputed issues not only in *Models of Doom*, but also in *Nature* [7], in *Simulation* [11], in a column in *The Observer* (where the backcasting argument is described as "trading insults like prize-fighters at a weigh-in") [9], and in *The New Republic* [2]. The present paper hopes to clarify the backcasting controversy by developing four points:

(1) Backward integration is an uncertain method of exercising dynamic models. The control mechanisms found in social systems are unstable when played backward. Numerically unstable backcasts may easily generate spurious results.

(2) But the Sussex team's backcast is accurate. The absurd 1880 population value shown on fig. 1 is not "completely spurious," but due to a misspecification in Forrester's model.

(3) Although backward integration of dynamic models may be numerically inaccurate, retrospective comparisons of model behavior with observation is practical. This may be accomplished directly by forward integration from pre-initial times.

(4) The pre-1900 error on World2 is not evoked by that model when it is exer-

cised after 1900. Hence, the retrodiction failure neither confirms nor disconfirms the validity of Forrester's conclusions. Further, simple modification allows World2 to "fit" 1880–1900 without changing the model's later behavior (or validity).

2. Backwards integration with dynamic models

Dynamic models, including the Forrester World Models, may be cast as sets of first order differential equations:

$$(\mathrm{d}/\mathrm{d}t)\boldsymbol{x}(t) = f(\boldsymbol{x}(t)) + \boldsymbol{u}(t) \qquad (1)$$
$$\boldsymbol{x}(t_0) = \boldsymbol{x}_0$$

where $\boldsymbol{x}$ represents the vector of system state variables, $\boldsymbol{u}$, the exogenous elements, and $\boldsymbol{x}_0$, the initial conditions.

In all cases of practical interest[2], system (1) has a unique solution through any initial condition. If a dynamic system is observable, then its starting point can be recovered. A (trivial) sufficient condition for observability is that all the elements of the system state vector be accessible [12, p. 1087]. Since, for computer simulation models, internal states are accessible, these models can be, in principle, integrated backwards.

For systems with analytically tractable solutions, future or past behavior are equally easy to find, but dynamic models of socials systems are usually complex enough so that closed-form analytic solutions do not exist or cannot be found. In this case numeric integration is used to recover approximate system trajectories.

Numeric integrations are corrupted by two sources of error – round-off error due to the finite capacity of digital computers, and truncation error arising from the particular integration scheme used. The importance of these errors for solution accuracy depends on the stability properties of the model being integrated.

Stability of dynamic models is very different in forward than in backward integration. These models are not only dynamic (differential equation) but *feedback dynamic*, consisting of clusters of positive and negative feedback loops. From the feedback loop structure comes asymmetry in backward vs. forward integration stability. Such feedback dynamic models have stable paths for forward evolution despite persistent, small perturbations of the scale of round-off and truncation errors, but have unstable paths for backward motion. Why is this true?

In any social system it is certain that perturbations appear due to exterior and adventitious actions. Ecological, economic, and managerial systems have evolved and survived or have been deliberately designed to maintain behavior

[2] A Lipschitz condition is sufficient; see [12, p. 1087].

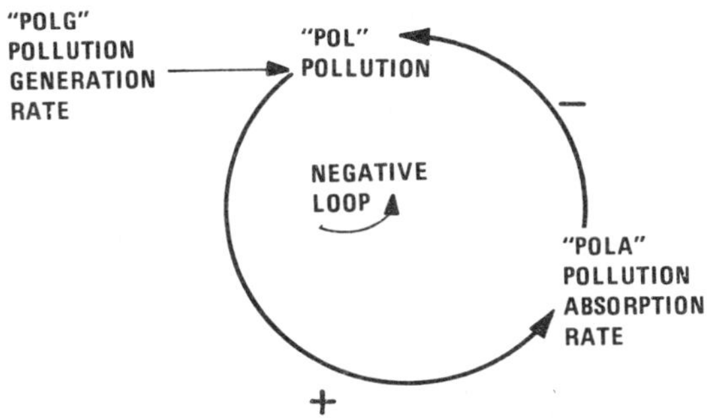

Fig. 2. The World2 pollution cleanup loop.

despite disturbances – certainly despite disturbances of the magnitude of those introduced by numeric integration. Indeed, any social system that could not survive small disturbances is no longer on the scene.

The element of structure that allows dynamic systems to tolerate small disturbances is the negative feedback loop. Negative loops are deviation compensating. They sense the discrepancy between a desired and an actual value for a state variable and initiate action to reduce the discrepancy, in forward integration, a negative loop is asymptotically stable. With the direction of time inverted, however, stable negative loops become unstable positive loops, loops that reinforce deviation.

An example is useful here. The element of Forrester's World2 that undermines backward integration is the negative feedback loop connecting pollution with the pollution absorption rate. This negative loop attempts to drive the level of pollution to zero. Simplified equations for the pollution loop in fig. 2 are[3]

$$
\begin{aligned}
&(\mathrm{d}/\mathrm{d}t)\,\mathrm{Pol}(t) = \mathrm{Polg}(t) - \mathrm{Pola}(t) + \xi(t) \\
&\mathrm{Pol}(0) \;= 0.54\mathrm{E}9 + \eta \qquad \text{(initial time is 1900)} \\
&\mathrm{Polg}(t) = (0.81\mathrm{E}9)\exp(0.04t) \\
&\mathrm{Pola}(t) = \mathrm{Pol}(t)/0.6.
\end{aligned}
\tag{2}
$$

Here, $\xi(t)$ and η represent the error introduced respectively by imperfections in numeric integration and in specification of initial values. Eqs. (2) have an analytic solution:

$$
\begin{aligned}
\mathrm{Pol}(t) = {} & \exp(-t/0.6)\,[\mathrm{Pol}(0) + \eta] \\
& + \exp(-t/0.6) \int_0^t \exp(s/0.6)\,[\mathrm{Polg}(s) + \xi(s)]\,\mathrm{d}s.
\end{aligned}
\tag{3}
$$

[3] Eq. (2) are a linearized form of Forrester's formulation accurate for the period 1920 to 1940 (or vice-versa). The original equations can be found in [6, p. 132].

Notice the $\exp(-t/0.6)$ terms. These terms show that during numeric integration small errors corrupting the initial value and the absorption and generation rates are rapidly forgotten. In order to accomplish a backward integration, we invert the sign of a model's differential equations. For the pollution subsector of eq. (2), in backward motion pollution is *increased* by absorption and *decreased* by pollution generation.

$$(\mathrm{d}/\mathrm{d}t)\,\mathrm{Pol}(t)_{\mathrm{backwards}} = -\mathrm{Polg}(t) + \mathrm{Pola}(t). \tag{4}$$

The solution to the backwards system is

$$\mathrm{Pol}(t) = \exp(+t/0.6)[\mathrm{Pol}(\text{“initial”}) + \eta] + \exp(t/0.6) \int_0^t \exp(-s/0.6)[\mathrm{Polg}(s) + \xi(s)]\,\mathrm{d}s. \tag{5}$$

Eq. (5) includes $\exp(+t/0.6)$ terms. Now small errors in initial condition specification and in the calculation of the rates of flow will be augmented over time. In fact, after ten years, the exponential term has a magnitude of e^{17}.

This example indicates the main source of instability in backward projection of dynamic models: negative loops played backward act as positive loops and compound integration error. Two questions arise at this point. First, since social systems include positive loops and positive loops are unstable, why then is there no difficulty in forward integration? The answer is simple: social systems include negative feedback loops that constrain explosive forward integration. No social system can grow by exp (17) in ten years; explosive growth will bring into dominance a compensating control loop. An epidemic process is at first dominated by a positive feedback loop. Any integration error is reinforced. However, as the epidemic progresses a significant number of infected are removed from a pool of those susceptible and infections decline. The system is eventually constrained by a negative loop. Integration error might shift the peak of the epidemic but would not change its general pattern.[4]

A second question asks: Are some numerical integration methods robust enough to accomplish accurate backward integration for dynamic models of social systems? Existing computational experience has not suggested a robust candidate. A paper by Wright [12, p. 1090] offers a series of examples illustrating unstable and inaccurate backward integration due to initial value errors and to errors of round-off and truncation, for both rectangular integration as well as for more sophisticated numerical methods. Backcasting instability is well known to control engineers. Burns and Malone report that optimal control algorithms which backward-

---

[4] An epidemic model backcasts easily; in backwards evolution it is dominated first by a positive loop then by a negative loop; exactly the same pattern that holds for its forward evolution. See [12, p. 1086].

integrate a model (or forward integrate its adjoint) ". . . generally meet with failure . . ." [3, p. 168] when applied to World2.

3. What happened in the world dynamics backcasts?

Fig. 1 shows two apparently similar backcasts of World2. The *Doom* backcast is valid; the *Limits* backcast, spurious. To differentiate inaccurate from accurate backward integrations we should examine the behavior of all the system's state variables and consider its nonlinear functions. For both of the runs of fig. 1, the pollution loop (fig. 2) explodes. For the *Doom* run, pollution approaches negative infinity. In the *Limits* run, pollution explodes in a positive direction.[5]

The difference is important because World2, like all realistic social system models, is *nonlinear.* Forrester's formulation has increasing positive pollution act to strongly increase deaths, decrease food production, etc. But small, zero, or negative pollution quantities have no effect on the system's operation. The *Limits* population backcast is spurious because its incorrect pollution calculation forces World2 to generate phantom deaths. In the *Doom* run, pollution sector integration error is decoupled from the system and cannot be blamed for the strange population pattern. And as shown below, the *Doom* backcast has revealed the legitimate 1880 predecessor values for World2's variables (except pollution.)

4. Implications of backward integration instability

A backcast test checks whether trajectories of a model integrated backwards in time are consistent with historical observation for the pre-initial period. This implies the "reasonable" pre-initial time trajectories are known. A poor match might be due to error in the model (as in the *Doom* backcast) or to integration instability (as in the *Limits* run). Fortunately, the results of a backcast may be accomplished easily by starting the system with reasonable pre-initial time estimates, running it forward, and checking that it ends "close" to its initial state. Fig. 3 illustrates three possible results from a retrospective validation test.

Path 1, a backward integration from initial value x_0 indicates an unreasonable trajectory for the pre-initial period. Path 1 may point to the legitimate predecessor state of the model, or it may be a spurious result. Paths 2 and 3 are possible outcomes of forward integration from the sought for pre-initial state x_{00}. Since forward integration is stable, a result like path 2 or 3 is nonspurious. A result like path 2 shows the model to behave reasonably for the pre-initial behavior. A result like path 3 shows unreasonable pre-initial behavior. In this last case, the model fails a retrospective test.

[5] See [12, p. 1090] or [5, p. 64].

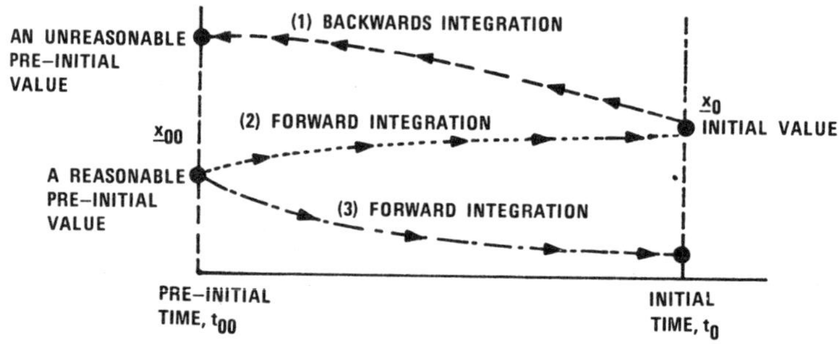

Fig. 3. Possible results from retrospective validation tests.

An example of path 2, a successful pre-initial test, is reported by Hamilton [8, p. 245]. A dynamic model of regional development was originally tested against 1960–1966 data. In a refinement phase, that model was re-initialized with 1950 values and shown to track during the 1950–1966 interval. The *World Dynamics* model offers an example of path 3: By forward integration, we may establish that World2's behavior is not consistent with any reasonable set of 1880 initial values.

World2 has five state variables: natural resources, pollution, capital fraction in agriculture, capital investment, and population. A historical 1880 population estimate is available: 1.45E9 people [1, p. 38]. Other reasonable 1880 values must obey common sense: less pollution, less capital, more natural resources. Forrester's equations imply that before 1900 resources, pollution, and agricultural capital have neutral impact on model behavior.[6] Population change, around 1900, is very sensitive to the value of "Material Living Standard, MSL", roughly, capital per person. If MSL is less than 0.25 (25% of its nominal 1970 value), population declines. An MSL specification high enough to grow population from its historical 1880 estimate to World2's 1900 starting value implies an 1880 capital value as great as Forrester shows it in 1900. Around 1900 the capital investment equation reduces to

$$(\mathrm{d}/\mathrm{d}t)\mathrm{Cap\ Inv} = (0.05)\mathrm{Pop} + (0.20)\mathrm{Cap\ Inv}. \tag{6}$$

Eq. (6) shows strong growth. If population is to grow properly, capital must stagnate. If capital evolves correctly, population declines.

In a word, something is seriously wrong with World2's specification for ranges of operation that hold before 1900. But what does this say about the conclusions

[6] The reader may check the assertion by writing linearized equations for Forrester's World2 around its 1900 values. Or see [12, p. 1091].

[7] "MSL" is a worldwide measure of Material Standard of Living, roughly equivalent to capital investment per person, normalized so that its 1970 value is unity. See [6, p. 36].

drawn by Forrester from his model? And in general what is implied by failure of a retrospective test?

5. Interpreting results of retrospective tests

Retrospective tests can play a useful role. In the Hamilton example above, a longer period of good fit increased client confidence. So long as "the same mechanisms operated in the past," extending the time horizon over which a model shows realistic behavior improves model credibility. Because social systems are nonlinear the "same mechanisms" may not have operated in the past. Drawing proper inferences from nonspurious backcast failures is nontrivial.

The *Doom* team asserts

> "the credibility of the futures portrayed by the [world] models rests on their ability to reproduce to a plausible degree historical trend between 1900 and 1970, and, *in general, the longer the period over which trend forecast agree with history the more valid they appear*" [5, 113, emphasis added].

This assertion is not correct. A linear model will in principal retrodict the behavior of its referent linear system behind the model's initial time provided values for exogenous inputs are known for the pre-initial period. Nonlinear models will retrodict referent nonlinear systems if exogenous inputs are available for the pre-initial period *and* if the model formulation is accurate for the range of operation that occurs during that earlier period.

A dynamic model is constructed to be adequate over a specific range of values of its state variables. If, in a time period earlier (or later) than that for which the model was designed, substantially different values of state variables hold, then the model cannot be expected to behave like its referent system. It is the responsibility of the modeler to indicate the range of applicability of his work. While Forrester did express concern for the plausibility of extreme points of some functions [6, p. 40], he did not offer a clear statement of World2's valid range. In fact, that model is not valid for levels of capital investment holding before 1900.

Peculiar 19th Century trajectories of World2 result from poor formulation of death rate and of capital investment rate as functions of "Material Standard of Living." Results of the pre-1900 tests of Forrester's model show these two functions to be in error for values of MSL below 25% of the 1970 worldwide Material Living Standard. Since, in *none* of the ten runs reported in *World Dynamics* does MSL fall below this 25% value, Forrester's conclusions are not refuted (nor supported) by this model's retrodiction failure. By luck then, the misformulation is not relevant.

Yet, the poor formulation may still be worrisome. It is the cause for strange behavior of world models when disaggregated into developed and underdeveloped regions [5, p. 121]. Population and capital are not equally divided between re-

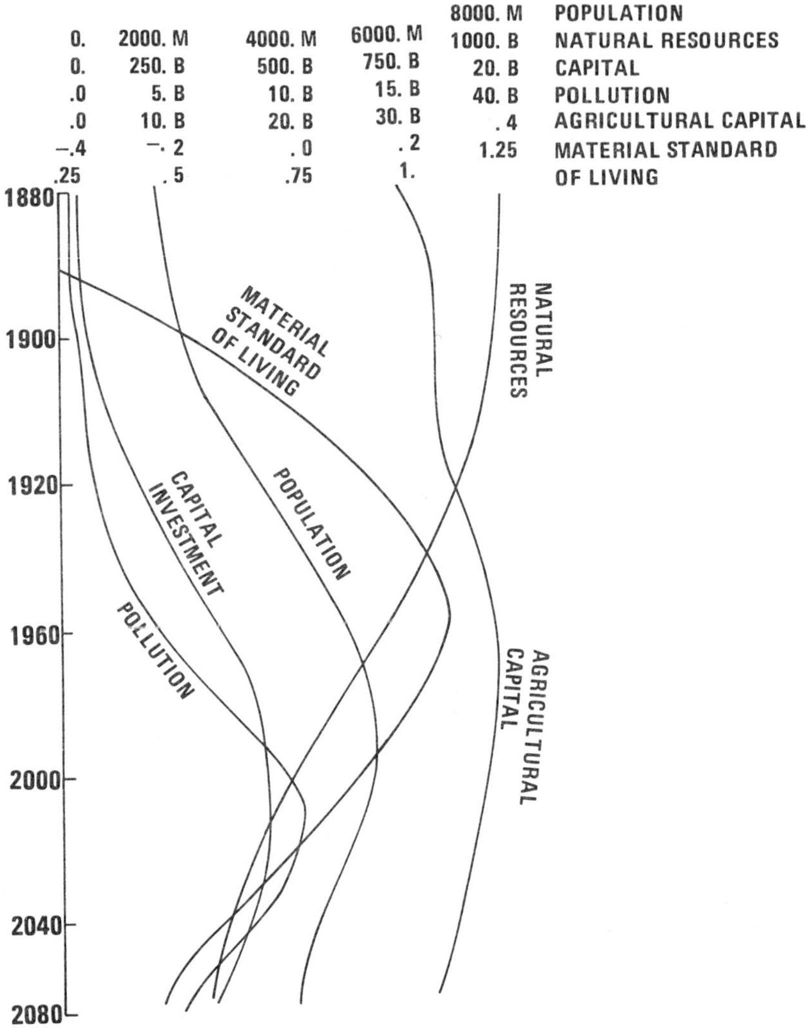

Fig. 4. Behavior of World2 with 1880 "patch".

gions, values for material living standards in the underdeveloped world are low enough to fall in the invalid range of death and capital investment rates. One would be more comfortable with a formulation valid for a wider range. Indeed, if the assertion is correct, if Forrester's misspecification falls in a range not evoked after 1900, then one should be able to "patch" the World2 formulation. This patch should be plausible, should result in reasonable behavior from 1880–1900, and have no effect on post-1900 results. The modifications below do extend the World2 model in the fashion desired:

Capital Investment Multiplier (CIM) is set to 0.16 if Material Standard of Living is less than 0.25 (in the original formulation CIM ranged from 0.1 to 0.325 as MSL ranged from 0 to 0.25).

Death Rate from Material Multiplier (DRMM) is set to 2.4 for MSL less than 0.25 (in Forrester's original run DRMM ranged from 3.0 to 2.4 as MSL ranged from 0 to 0.25).

1880 initial conditions are selected as: population = 1.45E9; natural resources = 900E9; capital investment = 0.4E9; pollution = 0, capital investment fraction in agriculture = 0.16.

Fig. 4 shows 1880–2100 behavior with the modifications.

These trajectories lie exactly on top of those in Forrester's World2 base run [6, p. 70]. Note that S curve – Material Living Standard, is below scale (0.25) from 1880 to 1904, but never again falls into this range.

This "patch", of course, does not confirm the adequacy of Forrester's model. It merely shows that World2 is not inconsistent with pre-1900 observation.

6. Summary

Validation testing by simulating dynamic models backwards in time is an uncertain scheme, open to spurious results. Feedback dynamic models when played backwards are unstable. Backward integration instability does not preclude retrospective validation testing: Pre-initial time period checks can be easily accomplished by forward integration. Proper evaluation of the results of retrospective tests depends on a model's nonlinear properties and range of application. The failure of World2 model to project accurately for 1880–1990 neither disconfirms nor confirms the conclusions Forrester drew from his model's post-1900 performance.

References

[1] Donald J. Bogue, Principles of Demography (John Wiley, New York, 1969).

[2] Kenneth Boulding, Zoom, gloom, doom, and room, The New Republic (August 11, 1973) 25–27.

[3] James R. Burns and David W. Malone, Optimization techniques applied to the Forrester model of the world, IEEE Transactions on Systems, Man and Cybernetics, SMC-4, No. 2 (March 1974).

[4] H.S.D. Cole and R.C. Curnow, Backcasting with the world dynamics models, Nature 243 (May 11, 1973).

[5] H.S.D. Cole and R.C. Curnow, Models of Doom (Universe Books, New York, 1973).

[6] Jay W. Forrester, World Dynamics (Wright-Allen, Cambridge, 1971).

[7] Dennis Gabor, Letter to Nature 243 (May 25, 1973) 249.

[8] H.R. Hamilton et al., Systems Simulation for Regional Analysis (M.I.T. Press, Cambridge, 1969).
[9] Nigel Hawkes, A clash of computers, The Observer (November 2, 1973).
[10] D.L. Meadows et al., The Limits to Growth (Universe Books, New York, 1972).
[11] Guy Streatfield, No limit to the growth debate, Simulation, 21, No. 6 (June 1973) p. 210.
[12] Richard D. Wright, Retrodictive tests of dynamic models, in: Proceedings 1973 Summer Simulation Conference, Montreal (Simulation Councils Inc., Box 2228, La Jolla, California).

BACKWARD INTEGRATION OF SYSTEM DYNAMICS MODELS – A USEFUL VALIDATION TEST?

Kenneth R. BRITTING

System Dynamics Group, Alfred P. Sloan School of Management, Massachusetts Institute of Technology

1. Introduction

There has recently been a flurry of debate over the use of retrodictive tests (i.e. running models backwards in time) as a measure of validity in system dynamics models. The essence of retrodictive testing is to simulate a model backwards in time to determine if the model indicates behavior trends which agree with historical data. For example retrodictive tests are used by geologists to determine if models of stalagmite formation are consistent with established geological theory.

The issue of retrodiction has been raised in the context of Forrester's World Dynamics model [1] by a research task force at the University of Sussex in Great Britain. In reporting on their comprehensive review of the mathematical and structural formulation of the World Dynamics model, the Sussex group states [2]:

> "... much of the credibility of the futures portrayed by the models rests on their ability to reproduce to a plausible degree historical trends between 1900 and 1970 and, in general, the longer the period over which trend forecasts agree with history, the more valid they appear. As it stands, World 2 [World Dynamics] will not run backwards from 1900 for more than a little over 20 years. ... [the retrodiction runs] seem to indicate that the 20th century lies in the aftermath of a catastrophic population collapse (from a previously infinite population) dated about 1880 with population still decreasing as late as 1904."

In a detailed response to the Sussex retrodictive tests, Wright [3] points out that the model flaw which results in the preposterous model behavior from 1900 to 1880 had already been acknowledged by Forrester [1, p. 80]. Moreover Wright was able to show that a slight modification of two constant model parameters would enable the model to retrodict in a manner consistent with historical data from 1900 back to 1880. The important point to be noted was the fact that behavior of the Wright-modified world dynamics model differs very little from the original model in the time frame for which the model was designed to operate, i.e. from 1900 to 2100. The fact that the model is rather insensitive to small changes in its parameter specification and initial conditions is a frequently encountered characteristic of nonlinear feedback models and the systems they represent.

Now that the debate on this specific issue has subsided due to its resolution, we should try to determine what the debate has contributed to our understanding of the broader issue of overall model validity. Since both the original model with its flaw and the corrected model show almost identical behavior after the year 1900, one is forced to conclude that the debate has shed more heat than light on the subject of validation. It is doubtful that many observers would conclude that the modified model is significantly more valid than the original model. True the range of model appropriateness has been extended from 1900 back to 1880, but in the process we have seen that the major conclusions, which are drawn from the model in the 2000 to 2100 time range, are virtually unaffected by its performance prior to 1900.

The purpose of this paper is to investigate the theoretical basis for the retrodiction of dynamic models and to explore the numerical difficulties that are encountered when attempts are made to retrodict using a digital computer.

2. Mathematical basis for retrodiction

System dynamics models fall into the class of ordinary differential equations which can be written in the form:

$$\dot{x}(t) = f[x(t), u(t)]; \qquad x(t_0) \text{ given} \tag{1}$$

where:

$x(t)$ = nth order vector of system states (or levels);
$u(t)$ = vector of exogenous inputs;
$x(t_0)$ = initial value for state vector at $t = t_0$;
$\dot{x}(t)$ = time derivative of the state vector;
$f(\ )$ = nonlinear vector function.

The set of equations representing a system dynamics model can always be manipulated into the form of (1).

2.1. Existence and uniqueness of solutions

For nonlinear systems described by (1), such basic properties as existence and uniqueness of solutions are not guaranteed. It is well known, however, that if the function $f[x(t), u(t)]$ satisfies a Lipschitz condition [4] there exists a unique solution passing through $x(t_0)$. In practical terms only functions containing infinite slopes or discontinuities will not be Lipschitz.

System dynamics model formulations frequently contain discontinuous functions. For example, logic functions which select either variable A or B depending on the value of a third variable C are frequently used. Moreover the simulation of policy intervention frequently requires that model parameters be discontinuously changed. However, within a domain of the function space where $\boldsymbol{f}(\ )$ is single-valued and continuous, the Lipschitz condition will be satisfied and a unique solution will exist passing through the initial condition $\boldsymbol{x}(t_0)$. In order to guarantee a unique solution to (1) it is then necessary to require only that the function $\boldsymbol{f}$ possess a finite number of discontinuities which are connected by segments of function space which are Lipschitz. The simulation model, i.e. the solution of (1), will then display a unique trajectory, guaranteeing that integration backward to $\boldsymbol{x}(t_0)$ will be along that same unique trajectory. Thus we conclude that the mathematical structure of system dynamics models will generally allow retrodictive tests to be validly performed. As will be pointed out in the next section, however, numerical difficulties in performing backward integration can produce spurious trajectories resulting in incorrect conclusions being drawn from the retrodictive tests.

3. Numerical difficulties in retrodiction

There are two possible sources of computational error which one can encounter in system dynamics simulations – round-off error and integration error. Round-off error is associated with the fact that the computer's word length is finite. Each time a computation is performed the computer must approximate the final digit. For the IBM 360 systems operating in the single precision mode the seventh significant digit is uncertain due to round-off. Although the effects of round-off error can be assessed using statistical analysis methods [5], its effects on the computer simulation will not be pursued herein since, in most applications, integration error is by far the more important effect.

Integration error is caused by the inability of the digital computer to perform continuous time integrations. The DYNAMO computer language, which is used for most system dynamics simulations, employs a very simple rectangular integration algorithm. We illustrate with a typical DYNAMO level equation:

$$X \cdot K = X \cdot J + (DT)(X \cdot J/\tau) \tag{2}$$

where $X \cdot K$ is the system state at time K; $X \cdot J$ the system state at the preceeding time step, J; DT the integration time step; and τ the system time constant. If in (2), $X \cdot J$ is moved to the left hand side and the equation is divided by DT, we see that in the limit as DT approaches zero, (2) represents the continuous system:

$$\dot{x}(t) = x(t)/\tau. \tag{3}$$

Thus, the solution obtained in the DYNAMO digital simulation (2) will approach the analytic solution of (3) in the limit as the time step, DT, approaches zero. The integration error can be made as small as we please by reducing the integration time step.

3.1. Integration error in linear first order systems

An analytic assessment of integration error resulting from rectangular integration can be obtained for simple first order linear systems. The results provide some insight into the propagation of error in more complex systems.

Consider a system which can be represented by (3), modified to describe either a positive or negative feedback system:

$$\dot{x}(t) = \pm\, x(t)/\tau; \qquad x(t_0) = x_0. \tag{4}$$

The plus sign applies for a positive feedback (exponential growth) system with the minus sign indicates a negative feedback (exponential goal-seeking) system. The analytic solution to (4) is given by:

$$x(t) = x_0 \exp(\pm\, t/\tau). \tag{5}$$

If the time, t, is divided into n intervals of length DT, $t = n(DT)$, the above exact solution after n time steps becomes:

$$x_n = x_0 \exp[\pm n(DT)/\tau]. \tag{6}$$

The consequences of solving (4) using a rectangular integration scheme can be assessed by applying the definition of the derivative to (4) and rearranging:

$$x'_{n+1} = (1 \pm DT/\tau)x'_n. \tag{7}$$

We have introduced the prime notation to distinguish between the exact solution of (6) and the rectangular integration rule approximation of (7). If the system is initialized at x_0, (7) yields:

$$\begin{aligned} x'_1 &= (1 \pm DT/\tau)x_0 \\ x'_2 &= (1 \pm DT/\tau)^2 x_0 \\ &\vdots \\ x'_n &= (1 \pm DT/\tau)^n x_0. \end{aligned} \tag{8}$$

A comparison of (6) with (8) reveals that for all $DT/\tau < 1$, rectangular integration underestimates the solution to both positive and negative first order systems. In effect the numerical solution will appear to lengthen the time constant for positive feedback systems and shorten it for negative feedback systems. This modification of the effective time constant becomes more pronounced as DT/τ approaches unity.

A quantitative grasp of the situation can be obtained if we define the equivalent time constant, τ', associated with the numerical solution as the time at which the solution reaches a value of: $x'_n = x_0 \exp(\pm 1)$. Using (8), it can be shown that the equivalent time constant is related to the actual time constant through the relationship:

$$\tau'/\tau = \pm (DT)/\tau)/\ln(1 \pm DT/\tau). \tag{9}$$

The above relationship is plotted in fig. 1. It is seen that the variation in the simulated system's time constant can be kept below about 5% if $DT/\tau < \frac{1}{10}$.

Although the effect of integration error appears to be rather modest for reasonable values of DT/τ, the errors are compounded if retrodiction is attempted since the error accrued in the forward direction is multiplied by the backward integration error. The compounding effect is illustrated by retrodicting a negative system as shown in fig. 2. For two values of DT/τ, the system is initialized at x_0 = 100, integrated forward in time for $t = 4.6\tau$, and backward to $t = 0$. When the numerically obtained trajectories are compared with the plotted analytic solution, the retrodiction error is seen to be 38% and 60% for the $DT/\tau = \frac{1}{10}$ and $\frac{1}{4}$ cases, respectively.

We have shown that because rectangular integration underestimates the true solution for forward or backward running systems, the error compounding effect can lead to significant retrodiction errors. We note that *the error accumulation is a mathematically well behaved phenomena and does not cause the solutions to explosively diverge.* In fact we have seen that the computed solution will always be less than the true solution. The limitation of integration error through a suitable choice of DT is a very simple matter – one merely reduces DT until there is no significant effect in the computer simulation.

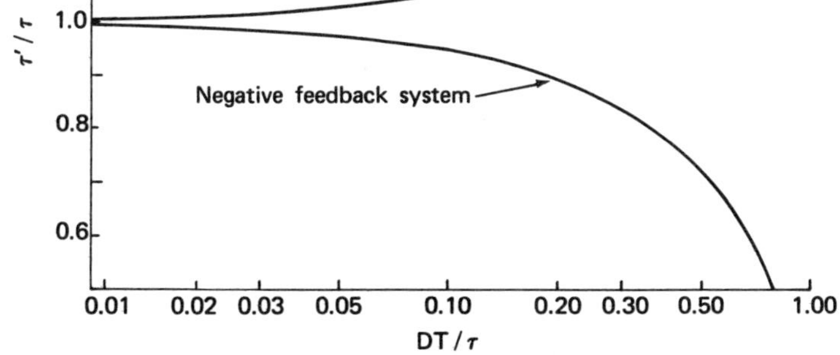

Fig. 1. Effect of numerical integration error on effective system time constant.

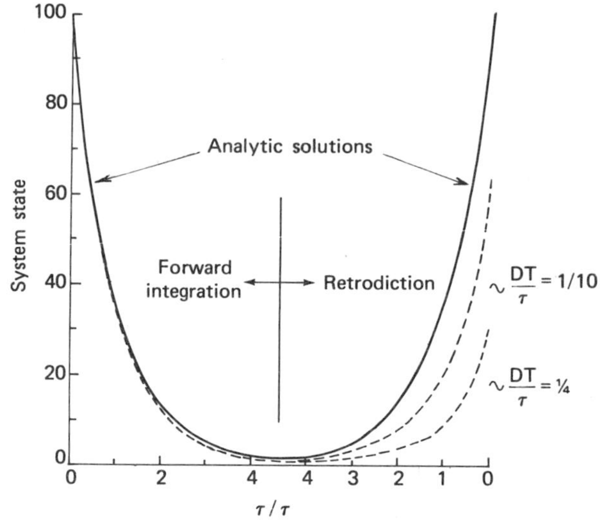

Fig. 2. Retrodiction of negative first order system for several values of D/τ.

3.2. Retrodiction in more complex systems

For complex, nonlinear systems the propagation of integration error is not as simple as the situation analyzed in the previous section. As will be demonstrated in this section, backward integration can lead to explosive growth for systems possessing certain mathematical properties.

Consider a simplified social system model which describes the interaction of population and housing in an urban area [6]. The system is described by the nonlinear differential equation:

$$\dot{P}=P\left(B-\beta\frac{P}{H}\right),\quad P_0=30;\qquad \dot{H}=H\left(\alpha\frac{P}{H}-D\right),\quad H_0=10, \qquad (10)$$

where P and H are the population and number of housing units respectively. The parameters B, D, α, and β are constants. For a parameter choice of $\alpha=\beta=1, B=4$ $D=2$, and $P_0/H_0=3$ the analytic solution to (10) is: $P(t)=P_0\exp(t)$ and $H(t)=H_0\exp(t)$. A computer simulation of (10) using rectangular integration [3, p. 1088] has demonstrated reasonable behavior with regard to integration error; i.e. the simulation converges to the analytic solution as the integration time step is reduced.

If the system is retrodicted, i.e. time reversed, (10) becomes:

$$\dot{P}=-\dot{P}\left(B-\beta\frac{P}{H}\right);\qquad \dot{H}=-H\left(\alpha\frac{P}{H}-D\right). \qquad (11)$$

The analytic solution for the same parameter choices as used previously are:

$P(t) = P_0 \exp(-t)$ and $H(t) = H_0 \exp(-t)$. Thus the system should retrodict backwards along the forward paths. However, the computer simulations show retrodiction trajectories which explosively diverge for any choice of time step. Evidently this particular model, although possessing a well behaved analytic solution under retrodiction, is extremely sensitive to small computation errors.

Some insights into the behavior of this system can be obtained through a linearization of (10) and (11). We examine small deviations from the analytic solutions by letting: $P(t) = P_0 + \delta P$ and $H(t) = H_0 + \delta H$, where δP and δH are perturbations from the trajectories of $P(t)$ and $H(t)$, respectively. If the perturbed variables are substituted into (10) and (11) and terms containing powers of the perturbation variables are assumed to be small, sets of linear differential equations emerge. For the forward integration case, (10) yields:

$$\delta \dot{P} + 2\delta P - 9\delta H = 30 \tag{12a}$$

$$\delta \dot{H} + 2\delta H - \delta P = 10. \tag{12b}$$

The characteristic equation [4, p. 76] associated with the above system of equations is:

$$(p + 5)(p - 1) = 0. \tag{13}$$

In the above equation, p is the derivative operator, $p = \mathrm{d}/\mathrm{d}t$.
For the retrodiction case we obtain the following set of linear equations:

$$\delta \dot{P} - 2\delta P + 9\delta H = -30 \tag{14a}$$

$$\delta \dot{H} - 2\delta H + \delta P = -10 \tag{14b}$$

with the characteristic equation:

$$(p - 5)(p + 1) = 0. \tag{15}$$

An examination of (13) and (15) reveals a somewhat suspicious situation. For the forward integration case the presence of the growth mode, the $(p - 1)$ factor, is expected since the system exhibits exponential growth. On the other hand, in the retrodiction case the $(p - 5)$ growth mode is not expected since the analytic solution indicates exponential decay. As will be demonstrated in subsequent paragraphs, the presence of the growth mode in the retrodiction equations causes the numerical solution to explosively diverge from the analytic solution.

For the forward integration case, (12) can be solved for the transfer function which relates population to its initial value:

$$\delta P = \frac{30(p + 5)}{(p + 5)(p - 1)}. \tag{16}$$

Similarly for the retrodiction case (14) yields:

$$\delta P = \frac{-30(p-5)}{(p-5)(p+1)}. \tag{17}$$

A comparison of (16) and (17) reveals why this system will not retrodict. Both transfer functions contain a mode cancellation. In the forward case, the $(p + 5)$ decay mode cancels leaving the exponential growth mode corresponding to the analytic solution. When the system is retrodicted the $(p - 5)$ growth mode cancels giving the system the exponential decay characteristic of the analytic solution. When the equations are numerically integrated, computational inaccuracies prevent the cancellation from actually occuring. In the forward case the lack of cancellation of the decay mode leads to little loss in precision since the growth mode completely dominates the solution. However, when the system is retrodicted the inexact cancellation of the growth mode leads to an exponential amplification of the computation errors which overwhelms the desired exponential decay behavior.

In summary, we have shown that it is possible to determine if a system will have difficulty retrodicting. In each special situation, we must determine the effect of an imprecise mode cancellation on the system's behavior. Performing the necessary analysis requires a great deal of effort but is absolutely necessary if valid conclusions are to be drawn from retrodictive testing. Forrester's World Dynamics model evidently contains mode cancellations similar to the example treated herein since a number of workers have reported on the inability of the model to retrodict [2, p. 80; 7]. Because of the nonlinear character of the world model, the system modes change as the system passes from growth through a transition phase to equilibrium. Thus it is likely that the model's susceptability to spurious behavior during retrodiction will be highly variable.

4. Conclusion

This paper argues that although retrodictive tests of complex nonlinear models is theoretically possible, their execution can possibly lead to spurious results due to computational error.

The broader issue concerns the question of what one learns from retrodictive testing. A properly formulated model is constructed to be "valid" within a particular time span of interest. The model builder seeks a model structure, i.e. a choice of relationships and parameters, which is appropriate to the model purpose over this fixed period of time. Retrodictive tests draw conclusions from the model's behavior outside of its design range. The model builder is under no obligation to ensure reasonable behavior outside of the model's time span. As models are constructed to encompass longer time spans, the structure must generally be modified to account for the changing relationships between the variables and parameters as the system evolves with time. Thus it is clear that retrodictive testing cannot be a meaningful procedure for assessing the validity of models.

References

[1] J.W. Forrester, World Dynamics (Wright-Allen, Cambridge, Mass., 1971).
[2] H.S.D. Cole et al., Models of Doom (Universe Books, New York, 1973).
[3] R.D. Wright, Retrodictive tests of dynamic models, in: Proceedings of Summer Computer Simulation Conference, Montreal, July, 1973.
[4] K. Ogata, State Space Analysis of Control Systems (Prentice-Hall, New York, 1967).
[5] A. Monroe, Digital Processes for Sampled Data Control Systems (Wiley, New York, 1962).
[6] W.I. Wils and P.M. Senge, Forward convergent systems, MIT System Dynamics Memo D-1809-1 (June 1973).
[7] J.R. Burns and D.W. Malone, Optimization techniques applied to the Forrester model of the world, IEEE Trans. on Systems, Man, and Cybernetics SMC-4, No. 2 (March 1974).

GLOBAL METHODOLOGIES AND LIMITED ANALYTIC TECHNIQUES – A PROPOSED MERGER TO ATTACK COMPLEX PROBLEMS

Rolf H. CLARK
School of Business Administration,
University of Massachusetts

Between the global methodology of the type underlying *Limits to Growth* and the more limited techniques such as inventory theory, there is much room for analysts to explore new ground. The point of this paper is that the global and limited approaches can often be merged to attack problems which neither is suited for individually. Two examples of blending such methodologies are presented: The first discusses interaction between decision theory and systems dynamics; the second concerns a merging of systems dynamics and linear programming.

In cursory terms, decision theory is concerned with making current decisions based on expected future states of the world. But expectations of future states may depend on future decisions. This feedback situation then means that current decisions depend on future decisions. Systems dynamics may be better suited to handle such situations than are other techniques which require rather limiting assumptions. Systems dynamics simulation can predict future decisions much as it can predict future states of the world. To do so the dynamic model functions within the constraints of prespecified decision strategies.

An example is appropriate.[1] In a recent analysis of military manpower planning a problem was encountered in which the manpower supply/demand gap over time depended on the type of weapon systems accumulated. Projecting long term manpower deficiencies required knowing what future decisions would be made on the acquisition of systems which had not yet been designed or even conceptualized. In order to obtain a reasonable solution, a feedback simulation model was developed which credited future decision makers with the following rationale: no matter what alternative systems were available, the acquisition process would take into account the relative scarcity, as measured by opportunity costs, of different manpower types. For example, if skilled manpower was in scarce supply, systems accumulated would be biased toward those requiring relatively unskilled persons. "Systems" were necessarily defined in the aggregate sense, and the mar-

[1] Both applications presented below were encountered in a Navy manpower research project done in conjunction with Robert A. Comerford.

ginal utility approach generated future decisions which were economically efficient.

The decision generator in this case depended on a bivariate density function, with one dimension measuring the total manpower intensity (of weapon systems to be acquired) and the other measuring skilled manpower intensity. The bivariate distribution was sampled from each time the model predicted investment funds would become available. Thus if X dollars were to be expended, the sample results would estimate the total manpower increment and skilled manpower increment of the systems obtained with the X dollars. These increments were added to the current manpower demand levels, and demand then compared with supply (estimated elsewhere in the model). The parameters of the bivariate sampling distribution were affected by the demand/supply gap existing when an expenditure occurred. In essence, if demand exceeded supply, the parameters biased the marginal distributions so that weapon acquisition tended to remove the demand/supply gap. In short, by imbedding a decision policy within a systems dynamics model, future decisions on the acquisition of weapons not yet conceptualized were estimated. This allowed considering planning horizons far longer than would be possible if only visualized system alternatives were considered.

A second example of merging global and limited methodologies concerns the duality of linear programming and the visibility of systems dynamics output. The dual variables, it is well known, provide the relative worths of marginal increases in restricted resources. Given some flexibility (over time) in resource development, the duals can indicate how this flexibility should be exercised – e.g., should some of resource A be sacrificed at time t_i so that more of B is available at t_j? How much should be sacrificed? In their purest form such deliberations fall under the purview of optimal control theory. But the phasing relationships between reallocations and their lagged effects may be too complex to find an optimum, perhaps because of discontinuities in underlying functional relationships. A systems dynamics model which parallels the system modelled by the linear program may prove helpful.

In another manpower application, manpower demand/supply gaps over time were needed for sensitivity considerations. The magnitudes of the gaps were studied relative to the rate of change of the magnitude – a process implicit in control theory solutions. Some indication of the time lags between resource reallocations and their effects was needed. These lags would indicate when reallocations should be implemented, that is how the time-phased trade-off between scarce resources should be coordinated. A systems dynamics model and a linear program were independently developed to model the situation being studied. The descriptive systems dynamics model yielded a printout of the pertinent time lags, and the time lags then allowed adjusting a heuristic which reallocated resources in the normative linear program. Using the heuristic new time-indexed resource constraints ("right-hand sides") were obtained and new iterations of the linear program run. Summarizing, a problem not tractible using the more theoretically correct calculus of variations was satisfied by an analytic marriage of convenience. Specifically, using a systems dynamics model to influence an adaptive linear program provided an acceptable solution.

Both the applications presented involved sacrificing theoretical purity for the sake of acceptable, if not optimal, solutions. But the cost of analyzing real-world complexity is to sometimes forego the cherished neatness of our most analytically perfect methods. Perhaps cross-breeding the purity of our optimizing methods with the vigor of techniques such as used in *Limits to Growth* will yield a healthy analytic strain. For then not only will problems be attacked, but the no-man's land between theoretician and practicioner will be cultivated.

LIMITS TO MODELS

Michael H. ROTHKOPF
Xerox Palo Alto Research Center

This paper examines the limits of what can be achieved by formal world modeling efforts. It focuses attention on the choices between use of resources in modeling efforts and in alternative uses, between the construction of a total system model and of partial system models, and between the construction of a multipurpose model and of multiple models. Above all, this paper urges recognition of the limitations of the fruits of modeling efforts.

World modeling, like most modeling efforts, is normally undertaken as means to achieving better decisions. There are limits to what even the most heroic modeling efforts can accomplish. This paper examines some of the limitations of world models and some of the strategic choices these limitations force on those who contemplate model building as a means of improving global policies.

While modeling may sometimes be helpful in the analysis of global issues, persons who are concerned with these issues and are considering using models in analyzing them, must first face some tactical questions. First of all, potential modelers must be realistic about what can be achieved with a modeling effort. They must consider that even with effectively unlimited resources there are limits to what they can hope to achieve. With the resources they can actually hope to muster, they are even more limited. They should recognize the tradeoff between the use of resources in building models and the use of resources in attacking the problem in non-analytic ways.

If this analysis leads to a belief that modeling is worthwhile, there is a choice to be made between building a total systems model of everything that affects the choice of alternative policies or of building a partial model. In Forrester's "total systems" philosophy for modeling social systems, it is better to include relevant but poorly understood phenomena than to leave them out. When this is done, however, the model is reduced to a model of the model builder's perception of the system. The opportunity for meaningful validation is greatly reduced. This is not necessarily bad provided the model builder avoids confusing this model of his opinions with a reliable model of the real system. In any event, the modeler should certainly consider the alternative of building a reliable model of part of the system. This might prove much more useful.

A model builder concerned with more than one policy issue must make another choice: should he build a multipurpose model or should he build multiple models? Model building is the art of abstraction. Full realization of this leads one away from confusing a model with reality and away from *general purpose* world models. An abstraction adequate for answering one question may be totally inappropriate for answering another. A model airplane that looks like a 747 will be quite different from a model airplane that flies like a 747. In model building, every approximation must checked for its neutrality with respect to each of the questions to be answered by the model. If the model is built before the questions are framed, it is likely to fail these checks. The more two questions differ and the harder they are to answer, the more difficult it becomes for a single model to answer both adequately. Maps are models, but we use maps with different projections, different scales, and different detailing for answering different questions.

Finally, it is extremely important to the effective use of modeling that completed models be carefully and realistically evaluated. All of the assumptions including omissions should be reviewed.

The results of the modeling effort can sometimes be used to establish the unreasonableness of the assumptions that went into the model. This should always be considered. If people believed the predictions of models, would they take action to make its predictions false? Does the model show the modeled system behaving in a way that was not anticipated when the assumptions were made? If so, are the assumptions, including omissions, still reasonable? For example, if the model shows a majority of the world's population dying off during a twenty year period because of high pollution levels, is it still reasonable to omit pollution control activity from the model?

Models like the world models that attempt to predict unusually far into the future should pay special attention to the validity of time extrapolation mechanisms. A weather model that is useful for predicting weather one day ahead will not be useful for making forecasts a month in advance. An econometric model that is useful for forecasting the economy a year in advance may do poorly on a 10 year forecast. When the time horizon increases by an order of magnitude the relevant phenomena are likely to change. In a 5 year economic forecast, technology might be held constant; in a 50 year forecast, this would be foolish. In a 50 year economic forecast, the known laws of physics might be taken as given; in a 500 year economic forecast, this might be foolish.

Mathematical models can be thought of as falling into a spectrum that is typified at one end by compelling models for which each element entering into the model is the result of rigorous scientific measurement. The other end of the spectrum is typified by models that can be regarded, at best, as a quantification of the untested opinions of their builders. If one chooses to build a model that depends upon guesses at the nature of vaguely understood mechanisms, then one must live with the fact that the results of the modeling effort will lie toward the latter end of the spectrum. Such models are not necessarily worthless. They may serve as a useful vehicle for communication between specialists. They may help a model builder develop his own thoughts and locate inconsistent sets of opinions. However, conclusions drawn from such models should never be thought of or presented as scientific or validated; nor should it be claimed that they are "[unlikely] to be substantially altered by further revisions".[1]

[1] This quotation is from the Introduction of the *Limits to Growth*.

THE NIGGLING AND THE GRAND: AN ASSESSMENT OF WORLD MODELING

C. West CHURCHMAN
University of California

This paper is based on the assumption that world modeling and the systems approach are extremely old ideas which have appeared in a wide variety of cultures and times. The theme is that our own age seems to be emphasizing one aspect, namely the details of the world system, and strangely to be ignoring the meaning of it all in terms of our destiny. I've used the mean English word "niggling" to describe the first: "demanding very minute work and small details", in other words, "fussy". For the other, I've used "grand": "magnificent, splendid . . . impressive; possessing and expressing moral or intellectual greatness."[1] "Grand" apparently goes back to a Swedish word meaning breast, and thence to well developed, then to the male simulation of a developed breast, the expanded chest, then to magnificent.

As co-editor of this series, I hasten to add that the distinction between the niggling and the grand is not intended to apply to any other paper herein, though of course it may, because present emphasis is overwhelmingly on details. It's my opinion that the major emphasis in the past, on the other hand, has been on the ultimate goals, i.e., on the grand.

I can easily illustrate the distinction by some of my own niggling. A great deal has been said about the quality of data needed for world modeling, but there has been entirely too little effort spent on the technical details of how the data should be collected and used.

I am referring throughout to models which are used to help us tell what policies we should adopt, and not to descriptive or predictive models per se. First of all, the coefficients in any model are estimates of real parameters, and hence the measure of performance of any policy is also an estimate; the niggling technical detail is the question of whether a "decent" estimate exists. The simplistic notion that plugging averages or totals into the model yields a decent estimate is technically naive, although almost universally assumed.

But there is another technical aspect of data for modeling that is far more serious. Many investigators do show a concern about the quality of the data from the point of view of the manner in which it was collected and its relevance to the problems the model is supposed to address. But there is a prior relevance question

[1] The Universal Dictionary of the English Language, H.C. Wyld (ed.), (Routledge, 1932).

which should be asked, namely, whether past data, no matter how well collected, are relevant. Consider the cost of a research program. The data might consist of salaries, equipment, rental space, etc. But salaries might be entirely irrelevant if, for example, the research personnel could have been engaged in some other project whose payoff would have been much higher. In this case, the cost is the lost opportunity and not the salaries. This much is fairly obvious. What is not so obvious is the implication that to estimate costs accurately one has to make a whole system judgment which consists of judging the opportunity space of the decision makers. In my experience, most modelers ignore this technical requirement. There is good reason for ignoring it, because most of the time we don't have the foggiest idea how to make such judgments. In other words, we don't know how to collect data for decision making models, world or noworld. Indeed, many researchers must sense the paradox of data collection: in order to collect the relevant data one has to comprehend the whole system, which is the aim of the model. In order to escape this paradox, one may assume that past data are adequate with respect to future planning, which assumption says that the past was well managed; a reduction to the absurd!

All this complaint is niggling enough, and no wonder researchers ignore its fussiness. If we don't know how to make estimates based on data, let's at least do what we know how to accomplish: build the models, and hope that some wiser age will discover the secret of how to collect the data. Indeed, perhaps our "first approximations," based on very inadequate data, will guide us in making better approximations, to better models, to better approximations. At this point, we are entering the domain of the grand. What aspect of the world we humans inhabit gives us any assurance that such attempts really do lead to successive approximations? Why do we say that the first world models are "first steps"? Obviously because we believe that there is a real destiny in our inquiry: at the beginning, terribly crude, but heroic, first steps, then a better and more secure step, and then –? The "then" is grand. We, we poor, inadequate first steppers, have faith in a more adequate future group of inquiring systems which will overcome the inadequacies of our feeble approximations.

If we scientists were not so afraid of myth, we could indeed defend ourselves in this manner, pointing out that the essential character of the hero's adventures is that the hero has no a priori assurance of success – if he had he wouldn't be a hero. Perhaps we and our mathematics and computers are not so thrilling as the bare chested hero of yore armed only with axe or sword. But that's just our own age's inability to recognize the mystical quality of our instruments of battle.

Of course, many inquirers don't want to see themselves as heroes, because, among other things, if they do, they give up their right to be qualified experts with technically sound methods, i.e., they give up their right to niggle.

But let's pursue the grand for awhile. In the great tradition of the systems approach, or rationalism, the grand was always central. Justice, for example, is a grand idea; it stands for a pinnacle of human endeavor – that humans should live in a social environment which is just to all – no exploitation, no corruption, no power domination. It was Plato's dream in the *Republic.* You are not to be surprised that the nigglers have tried to nibble at its foundation, pointing out that the philosopher-king bears remarkable similarity to an enlightened fascist. Niggling always follows the grand idea.

The labels of the grand are many; to mention a few, wisdom and enlightenment (*I Ching*), biological fulfillment (Aristotle), rationality and understanding (Stoics and Spinoza), God, Freedom and Immortality (Kant). The last refers to one of the most common portrayals of the grand: the whole system is, or is subservient to, the most magnificent Being of all, and our human destiny is our worship or our obedience. This idea of the grand appears in many cultures, often in very elaborate patterns, as in the Aztec's models of their Gods.

No one can completely ignore the grand in world modeling, else the model is either a description of our inevitable destiny or a fabrication of someone's imagination. But I'm interested in world modeling that tries to tell us what we should do. Even that most niggling modeling of all, incrementalism, must assume some sense of direction from worse to better, unless the incrementalists are teasing us.

But we seem to have gradations of grandness. The minimal seems to be survival of the human species. This is not very grand; it's something like saying that we should always have some reserves in the savings account, no matter how badly we conduct ourselves. Or the niggling modeler may say, "look, I don't know what the destiny of the human race really is, because no two philosophers seem to agree; but whatever it is, other than racial suicide, we must survive and I'm helping build world models to keep us alive." There is a niggling reply to this piece of modesty, namely whether one can separate the survival-maintaining system from the rest of the system, which is a perfectly decent systems question. But the grander question is whether survival means anything magnificent if, for example, to increase its probability we must slaughter a beautiful species of bird, level a mountain or despoil the ocean? *Why* should we survive if we are *the* evil species of nature?

A little better, but still very modest, is the ethical idea that we should take no irreversible steps. This is certainly not very grand; the cowardly hero always keeps the home in sight so that if too fierce a monster appears, he can run home safely. Here again is the niggling question of how we know that a change is reversible. From the point of view of a thrush whose habitat is ruined by a development, it's of little comfort to say that we can bring back his species to some habitat; *his* death is irreversible. The grander question is whether the story of a world of reversibles isn't pretty dull stuff.

Survival and reversibility are attractive to world model builders because, in their own words, they can be objective and apply sound scientific methods, or in the words of this essay, they can niggle endlessly. Another attractive ethical idea which is perhaps grander, is what I call an "enabling" ethics, of which economic development is a popular example. The idea is that we struggle to create a world in which each of us can gain what he needs and desires. The grandness of this idea of human destiny becomes rather diminished if one tries to use Gross National or International Product as a measure of performance, for this obscures the central idea of equity. Some notion of the equitable distribution of power and maximization of cooperation seems essential in this version of human destiny. I think E.A. Singer caught the idea best of all, because he puts a dramatic twist into the story. Progress, he says, includes increase in economic advantage, science, technology and cooperation; but it also includes the hero to whom all stages of progress are "failures", requiring a breaking away from stability and success towards an unknown but hopefully higher stage. If instead of "heroic" one writes "revolutionary", then the story includes all those grand revolutionaries of the past and the future forever. Revolution – not necessarily violent – becomes an integral part of the grand design.

One of the remarkable features of this grand story is that it is not hierarchical. To explain what I mean, let's display some grand schemes based on hierarchy. Aristotle seems to be an excellent start. If, as he suggests, we could describe the hierarchy of living things, then we could understand our own place in the total destiny of living forms. Not surprisingly, once the idea occurs, he rather readily ascends the biological steps from nutrition to contemplation. Others over the ages have expressed the same idea in many variants: increased consciousness or increased complexity being examples. It is interesting to see how often the number seven appears from the earlier Hindu stages of spiritual development to the present. Another grand interpreter of the human psyche, C.G. Jung, might very well be able to explain the archetypal appeal of seven. Sometimes hierarchies are used as convenient ways of classifying living forms, as, say, in J. Miller's work, which has seven levels of living forms. But the grand conception of hierarchy indicates the journey of our destiny, to ascent from level to level, often leaving some tired traveler behind to live out his years at lower places.

Two examples of the supreme level will suffice: one is union with God, either in His Heaven or through simati; the other – which may well be the same thing – is the full development of consciousness. In many writers, both are attainable, and the curious will want to know what lies beyond. If nothing, the grand voyage ends in a sardonic joke: the mountain peak, having no beyond, is no mountain at all, an ateliological mystery. One climbs in order not to climb. Mercy, at least, might tell us to go back down and help others up the way. This point in the hierarchical story is complicated and need not be pressed further here, except to say that the "beyond" is likely to be ineffable and hence beyond the kind of discourse of this essay.

Singer's grand vision, I said, was not hierarchical, though I may myself be going beyond his intentions. The hero of Singer's journey does not feel he has reached some level, so that he sees below him the path he has traveled and above him what remains. In that awesome moment when he reflects on what has been done compared to what needs doing, he can only feel tragedy – not just of failure of deed, but of failure itself. Then comes tomorrow's rebirth after the hero's death, and the "comedy" of new ambition, love, friendship, faith occurs. One could make an eternal cycle of this grand voyage, as the artists have. But an equally grand conception is to recall idealism; in the hero's mood of comedy, he sees that the struggle has been towards the ideals of truth, good (meaning plenty but also love), and Beauty (meaning he knows not what, but at least the sense of joy and being). In such a mood, he may even build world models and measure the degree of attainment towards these ideals: how much we know of our world, how plentiful are our resources, how cooperatively we live, appreciating beauty as an illumination of the glory of our grand progress. Then he dies again in the deception of it all; he is a child fighting monsters in his fantasies, which these days he calls world models, knowing that the monsters of greed, politics, bureaucracy, and moral rage are far beyond his powers.

If it sounds in the end as though the grand needs to go far beyond operational definitions to portray its real meaning, it sounds well. The grand contains in its busom all we are as human or living beings, including the humorous. Every niggling model builder who one day suddenly senses the grandness of his enterprise, must, if he is to remain sane, eventually appreciate its ridiculousness. And does the matter end there – in a joke? Of course not. Humor is a handmaiden of the serious and the grand, from whose busom springs our hope and eternal source of nourishment.